CT-6-2

FOURTH EDITION

AF737789

CLASSROOM TRAINING HANDBOOK

NONDESTRUCTIVE TESTING

liquid penetrant

GENERAL DYNAMICS

Convair Division

PREFACE

Classroom Training Handbook - Liquid Penetrant Testing, CT-6-2, is one of a series of training handbooks designed for use in the classroom or as a reference book during on-the-job training of nondestructive testing personnel.

This book can be most effectively used by those persons who have successfully completed Programmed Instruction Handbook, PI-4-2, Liquid Penetrant Testing. In general, the Classroom Training Handbooks are a continuation of the material taught in the corresponding Programmed Instruction Handbooks, deleting some of the process details and adding tabular matter and other technical details that are necessary for the continued upgrading of professional nondestructive testing personnel.

Other Classroom Training Handbooks in the series include:

CT-6-3 Magnetic Particle Testing

CT-6-4 Ultrasonic Testing

CT-6-5 Eddy Current Testing

CT-6-6 Radiographic Testing

ACKNOWLEDGMENTS

This handbook was originally prepared by the Convair Division of General Dynamics Corporation under a joint arrangement with NASA's George C. Marshall Space Flight Center. Convair's activities in the preparation of nondestructive testing training materials were greatly enhanced and accelerated by the MSFC technical and financial participation. Quality and Reliability Assurance Laboratory personnel at NASA's MSFC were to a large degree responsible for the successful completion of that program. Their understanding of the problems involved in teaching difficult subject matter, their realistic handling of NASA agency reviews, and their efficient transmittal of reviewer comments, made the publisher's task simpler than it might have been. Convair considers itself fortunate to have been associated with NASA on that project.

Additional assistance in the form of process data, technical reviews, and technical advice was provided by a great many companies and individuals. The following listing is an attempt to acknowledge this assistance and to express our gratitude for the high degree of interest exhibited by the firms, their representatives, and other individuals, many of whom gave considerable time and effort to the project.

Aerojet-General Corp.; Automation Industries, Inc., Sperry Products Division; Avco Corporation; The Boeing Company; Grumman Aerospace Corp.; Lockheed Aircraft Corp.; Magnaflux Corp.; Martin Marietta Aerospace, Denver Division; McDonnell Douglas Corp.; Met-L-Check Co.; Rockwell International, North American Aerospace Group; Rohr Industries, Inc.; Shannon Luminous Materials Co., Tracer-Tech Division; St. Louis Testing Laboratories, Inc.; Turco Products Division, Purex Corp., Ltd.; Uresco, Inc.; X-Ray Products Corp.

CLASSROOM TRAINING HANDBOOK
LIQUID PENETRANT TESTING

CHAPTER 1: INTRODUCTION

TABLE OF CONTENTS

CHAPTER 1: INTRODUCTION

100 GENERAL

The complexity and expense of today's machines, equipment, and tools dictate the use of fabrication and testing procedures that will ensure maximum reliability. Nondestructive testing (testing without destroying) provides many of these procedures. Of the number of nondestructive testing procedures available, liquid penetrant testing — of which this handbook is concerned — is one of the most widely used.

101 PURPOSE

The purpose of this handbook is to provide the fundamental knowledge of liquid penetrant testing required by quality assurance and test personnel to enable them to: 1) ascertain that the proper test technique, or combination of techniques, is used to assure the quality of the finished product; 2) interpret, evaluate, and make a sound decision as to the results of the test; and 3) recognize those areas of exhibiting doubtful test results that require either retest or assistance in interpretation and evaluation.

102 DESCRIPTION OF CONTENTS

1. Arrangement

The material contained in this handbook is presented in a logical sequence and consists of:

- a. Chapter 1: Introduction and testing philosophy.
- b. Chapter 2: Liquid penetrant testing principles with description of procedures, applications, and capabilities.
- c. Chapter 3: Equipment and materials.
- d. Chapter 4: Testing techniques including selection of penetrant materials and processes.
- e. Chapter 5. Interpretation of test results with description of indications and their characteristics.
- f. Chapter 6: Quality control of penetrant materials.
- g. Chapter 7: Comparison and selection of NDT processes as related to the five nondestructive testing methods.
- h. Chapter 8: Safety precautions for fire and toxic hazards.

i. Appendix A: Glossary of terms relating to liquid penetrant testing.

j. Appendix B: Measurement conversion charts (U.S. customary to International System (SI) units).

2. Locators

At the front of each chapter is a table of contents referencing the major paragraphs in that chapter. Also included is a list of figures and tables, where applicable.

103 INDUSTRIAL APPLICATIONS OF LIQUID PENETRANT TESTING

Because of the basic characteristics of liquid penetrant testing, it is used to test a variety of both metallic and nonmetallic materials such as welds, forgings, castings, plastics, and ceramics, etc. Since liquid penetrant testing is capable of economically revealing surface discontinuities (variations in material composition) in a variety of dissimilar materials, it is one of the most effective tools available to quality assurance personnel.

104 TESTING PHILOSOPHY

Nondestructive testing is used to assure maximum reliability of machines, equipment, and tools. To accomplish such reliability, test standards have been set and test results must meet these standards.

105 PERSONNEL

It is imperative that personnel responsible for liquid penetrant testing be trained and highly qualified with a technical understanding of the test equipment and materials, the item under test (specimen), and the test procedures. Quality assurance personnel must be equally qualified. To make optimum use of liquid penetrant testing, personnel conducting tests must continually keep abreast of new developments. There is no substitute for knowledge.

106 TESTING CRITERIA

Modern manufacturing procedures dictate that faulty articles be discovered as early in the manufacturing process as possible. This means that each item must be tested individually before it is required to perform in a subassembly and that each subassembly be tested before it is required to perform in an assembly, etc. This building-block approach requires that test processes be selected and test procedures be generated at the lowest level in the manufacturing process in order that the highest reliability may be obtained with lowest cost.

107 TEST PROCEDURES

Approved procedures for liquid penetrant testing are formulated from analysis of the test specimen or article, review of its past history, experience on like or similar specimens, and information available concerning discontinuities in similar articles. It is the responsibility of personnel conducting or checking tests to ensure that test procedures are adequately performed, and that the test objective is accomplished. Procedures found to be incorrect or inadequate must be brought to the attention of responsible supervision for correction and incorporation into a revised procedure.

108 TEST OBJECTIVE

1. The objective of liquid penetrant testing is to ensure maximum reliability by providing a means of:

   a. Obtaining a visual image related to a discontinuity on the surface of the specimen under test.

   b. Disclosing the nature of the discontinuity without impairing the material.

   c. Separating acceptable and unacceptable material in accordance with predetermined standards.

2. No test is successfully completed until an evaluation of the test results is made. Evaluation of test procedures and results requires understanding of the test objective as well as a knowledge of the material from which the test article is made. It also requires a knowledge of the manufacturing processes that were involved.

CHAPTER 2: PRINCIPLES

TABLE OF CONTENTS

LIST OF FIGURES

CHAPTER 2: PRINCIPLES

200 GENERAL

Liquid penetrant testing, a nondestructive means of locating and determining the severity of surface discontinuities in materials, is based upon capillarity. Capillarity, or capillary attraction, is the action by which the surface of a liquid, where it is in contact with a solid, is elevated or depressed. The materials, processes, and procedures used in liquid penetrant testing are designed to facilitate capillarity and to make the results of such action visible and capable of interpretation.

201 PHYSICS

1. General

The phenomenon of capillary action is one of the most important forces in nature. The rate and extent of the action associated with capillarity depends upon such factors as forces of cohesion and adhesion, surface tension, and viscosity. Capillarity can be observed when a plastic straw is inserted into a glass of water. When the straw is inserted, the water molecules enter the straw and begin to attract other nearby molecules, pulling them up the straw by cohesion. This process continues as the water rises higher and higher. The water continues to rise until the pull of surface tension is equalized. Cohesive forces prevent the water from falling back down the straw. Capillary action as applied in nondestructive testing is somewhat more complex, since various surface conditions hindering or assisting the action are encountered. Liquid penetrants in nondestructive testing have low tension and high capillarity. Capillary action is illustrated in Figure 2-1.

2. Application of Penetrant

In liquid penetrant testing, the liquid penetrant is applied to the surface of the specimen, and sufficient time is allowed for penetration into surface discontinuities. (See Figure 2-2.) If the discontinuity is small or narrow, as in a crack or pinhole, capillarity assists the penetration. When the opening is gross in nature, such as a tear, the liquid may be trapped when poured over the specimen.

3. Discontinuity Indications

After sufficient time has passed for the penetrant to enter the surface discontinuities, the excess surface penetrant is removed. The removal process clears the surface of the specimen but permits the penetrant in the

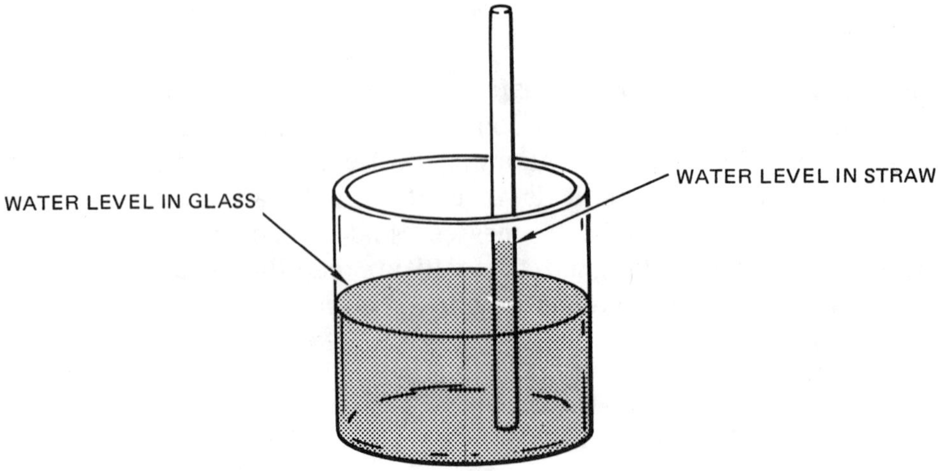

Figure 2-1. Capillary Action

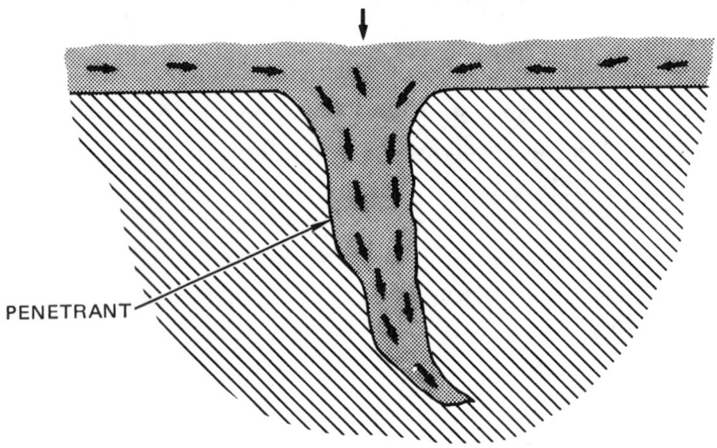

Figure 2-2. Penetration of Surface Discontinuities

discontinuities to remain. Capillary action is again employed in the process. A developer — which acts as a blotter — is applied to the test surface. (See Figure 2-3.) The blotting action of the developer draws the penetrant from the discontinuity and the penetrant appears on the surface of the specimen as an indication. The size of the indication, because of the diffusion of the penetrant in the developer, is usually larger than the discontinuity. There are also penetrants that provide sufficient discontinuity indication without the use of a developer; the developer is not required.

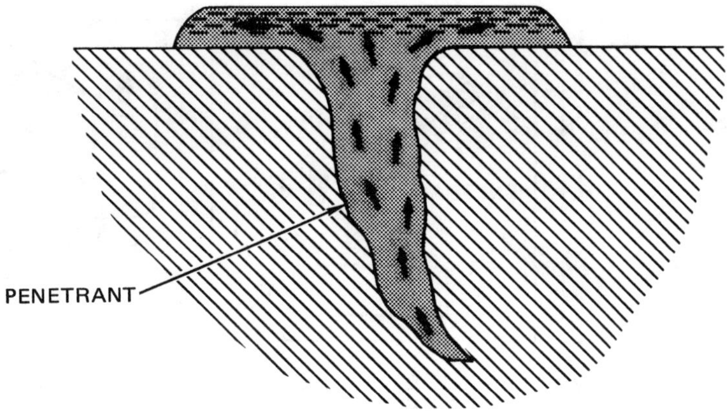

Figure 2-3. Reversed Capillary Action

202 VISIBILITY OF INDICATIONS

The ultimate success of liquid penetrant testing depends upon the visibility of indications. To ensure utmost visibility, the liquid penetrant contains either a colored dye easily seen in white light, or a fluorescent dye visible under black (ultraviolet) light. The dyes are obtainable in a variety of colors.

203 TEST PROCEDURE

The sequence of the test procedure, basically the same for all penetrant tests, can be broken into six main steps. These steps are illustrated in Figure 2-4, where it is shown that: 1) the surface of the specimen is first cleaned and allowed to dry; 2) penetrant is applied to the test surface and allowed sufficient time to seep into openings; 3) the penetrant remaining on the surface is removed without removing the penetrant from openings; 4) developer is applied to aid in drawing the penetrant back to the surface; 5) the surface of the specimen is visually examined to locate penetrant indications that have formed in the developer coating; and finally, 6) the test surface is again cleaned to remove developer residue and any minute traces of penetrant. The main steps and additional substeps required by the type of penetrant and process used are detailed in later chapters.

204 TEST PROCESSES

The penetrants used in liquid penetrant testing can be categorized by the type dye that they contain and the processing required to remove them from test articles.

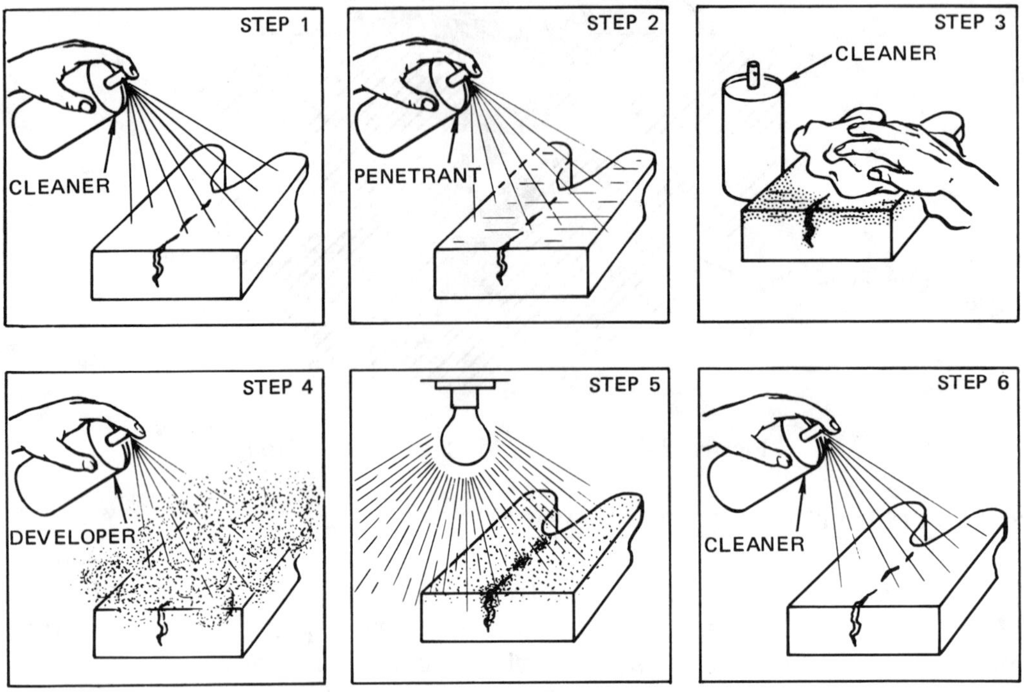

Figure 2-4. Sequence of Liquid Penetrant Tests

1. Types of Dye

Penetrants containing a visible color dye (usually red) are called visible dye, or color contrast, penetrants while those penetrants containing a fluorescent dye are referred to as fluorescent penetrants. Penetrants containing a combination of visible and fluorescent dyes are called dual sensitivity penetrants.

2. Processes

Processes employing penetrants that are self-emulsifying or removable with plain water are further classified as water-washable processes. Processes where a separate emulsifier is used to make the penetrant water washable are referred to as post-emulsified processes. And those processes in which the penetrant is removed by a solvent are identified as solvent-removed processes. Figure 2-5 illustrates the processing sequence used with visible dye and fluorescent penetrants; Figure 2-6 shows the processing sequence employed for dual sensitivity penetrants.

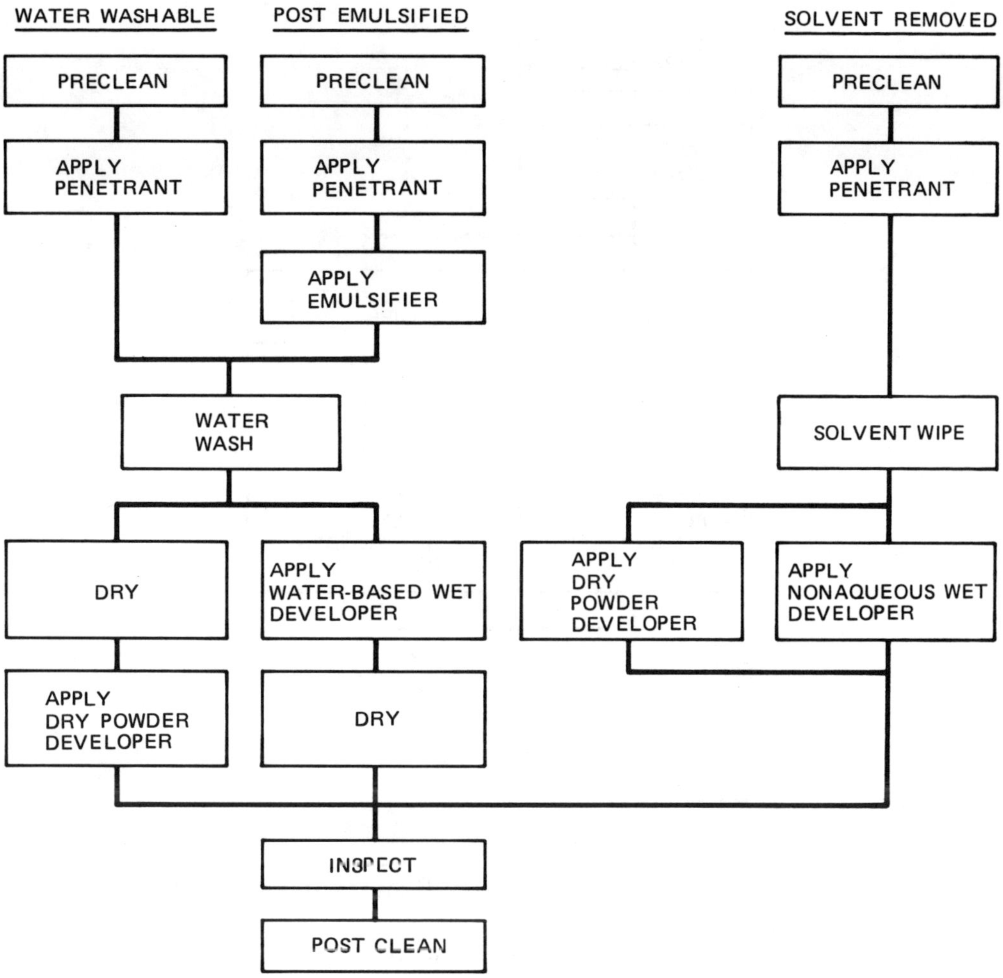

Figure 2-5. Visible Dye and Fluorescent Penetrant Processes

205 PROCESS SELECTION

Selection of the suitable penetrant type and process for a particular liquid penetrant test depends upon the sensitivity required; the number of articles to be tested; surface condition of the material under test; configuration of the test specimen; and availability of water, electricity, compressed air, suitable testing area, etc.

206 CAPABILITIES OF TEST

Liquid penetrant testing is capable of locating discontinuities open to the surface in articles made of any nonporous material. With penetrant tests, detectable discontinuities such as surface cracks, porosity, and "through" leaks can be found. These may be caused by fatigue cracks, shrinkage

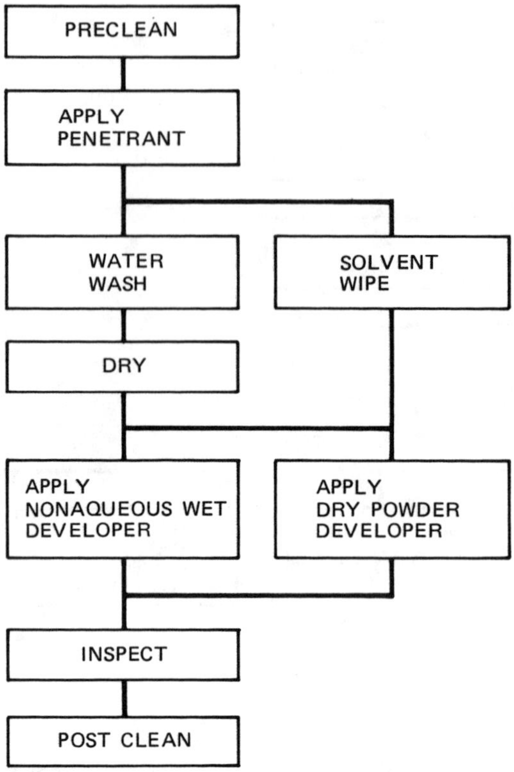

Figure 2-6. Dual Sensitivity Penetrant Processes

porosity, cold shuts, grinding and heat-treat cracks, seams, forging laps and bursts, as well as lack of bond between joined metals. Penetrant testing is successfully used on metals such as aluminum, magnesium, brass, copper, cast iron, stainless steel, and titanium. It also can be used to test other materials, including ceramics, plastics, molded rubber, powdered metal products, or glass. Since some plastics and rubber compositions are adversely affected by oil, sample tests are accomplished prior to actual testing of these materials to assure that the penetrant will not damage the material. Penetrant materials used in testing nickel alloys, certain stainless steels, and titanium must have a low sulfur and chlorine content. And materials used to test liquid oxygen (LOX) system components must be LOX compatible.

207 LIMITATIONS

The chief limitation of penetrant testing is that it can detect only those discontinuities that are open to the surface. It is also limited by its inability to test materials that have excessively porous surfaces.

CHAPTER 3: EQUIPMENT AND MATERIALS

TABLE OF CONTENTS

LIST OF FIGURES

CHAPTER 3: EQUIPMENT AND MATERIALS

300 GENERAL

The specific equipment and materials used in any liquid penetrant test are determined by the inherent requirements of the test procedure; the composition of the article under test; the size of the article; the frequency of like tests; and the size and type of suspected discontinuities. This chapter discusses the equipment and materials required to perform the various penetrant tests and the required precleaning and postcleaning.

301 PRECLEANING AND POSTCLEANING EQUIPMENT

1. General

Proper cleaning is essential to liquid penetrant testing for two reasons: 1) if the test article is not clean and dry, penetrant testing is ineffective; and 2) if all traces of penetrant test materials are not removed after test, they may have a harmful effect when the article is placed in service. The cleaning processes commonly used with penetrant testing are discussed in the following paragraphs. The equipment and material routinely used with these processes are all that are necessary for the cleaning required by penetrant testing.

2. Detergent Cleaning

Immersion tanks and detergent solutions are a common means of accomplishing the cleaning required by liquid penetrant tests. The detergents wet, penetrate, emulsify and saponify (change to soap) various soils. The only special equipment requirement imposed by penetrant test cleaning is the need for suitable rinsing and drying facilities. When thoroughly rinsed and dried, detergent cleaning leaves a test surface that is both physically and chemically clean.

3. Vapor Degreasing

Cleaning by vapor degreasing is particularly effective in the removal of oil, grease, and similar organic contamination. However, there are restrictions as to its use before and after liquid penetrant testing. Nickel alloys, certain stainless steels, and titanium havc an affinity for specific elements (e.g., sulfur or chlorine) and if exposed to them will become structurally damaged. Degreasing must be limited to those materials that have been approved for this method of cleaning.

4. Steam Cleaning

Steam cleaning equipment is particularly adaptable to the cleaning of large unwieldy articles not easily cleanable by immersion. No special equipment is required for steam cleaning of articles destined for liquid penetrant testing.

5. Solvent Cleaning

Solvent cleaning may use tanks for immersion, or the solvent material may be used in a wipe-on and wipe-off technique. Usually this cleaning process is used only when vapor degreasing, detergent cleaning, and steam cleaning equipment are not available.

6. Ultrasonic Cleaning

Ultrasonic agitation is often combined with solvent or detergent cleaning to improve cleaning efficiency and reduce cleaning time. The equipment is particularly useful in the cleaning of small articles.

7. Rust and Surface Scale Removal

Any good commercially available acid or alkaline rust remover may be used for precleaning. Required equipment and procedures are as specified in the manufacturer's directions.

8. Paint Removal

Dissolving type "hot tank" paint strippers and bond release or solvent paint strippers may be used to remove paint in precleaning. Required equipment and procedures are as specified in the manufacturer's directions.

9. Etching

Articles that have been ground or machined often require etching to prepare them for liquid penetrant testing. This process uses an acid or an alkaline solution to open up grinding burrs and remove metal from surface discontinuities. If an acid is used for etching, an alkaline solution is used as a neutralizing agent; if an alkali is used for etching, an acid is used as a neutralizing agent. The etching and neutralizing processes use either tanks and immersion or wipe-on and wipe-off equipment and materials.

10. Precleaning Processes To Be Avoided

Blast (shot, sand, grit, or pressure), liquid honing, emery cloth, wire brushes and metal scrapers should not be employed with liquid penetrant testing. These processes tend to close discontinuities by peening or cold working the surface. On occasion a wire brush may be helpful in removing rust, surface scale, or paint but it is used only when no other means of removal will suffice.

302 STATIONARY PENETRANT TEST EQUIPMENT

1. General

The stationary equipment used in liquid penetrant testing ranges from the simple to fully automatic systems and varies in size, layout, and arrangement depending on the requirements of specific tests. The size of the equipment used is largely dependent upon the size and types of articles to be tested. The layout of the equipment, i.e., whether a "U," "L," or straight line, is determined by the facilities available, the production rate, and the required ease of handling. The number of stations is dependent on the process used.

2. Stations

Depending on the type penetrant and processing employed (see Figures 2-5 and 2-6) the liquid penetrant test facility requires certain stations as shown in Figures 3-1 through 3-5. The required equipment components (stations) are combined to suit the particular test process. In a typical testing facility for a post-emulsification process, the following stations are required:

a. Precleaning Station (usually remote from penetrant test station).

b. Penetrant Station (tank).

c. Drain Station (used with penetrant tank).

d. Emulsifier Station (tank).

e. Rinse Station (tank).

f. Developer Station (tank).

g. Dryer Station (usually an oven type).

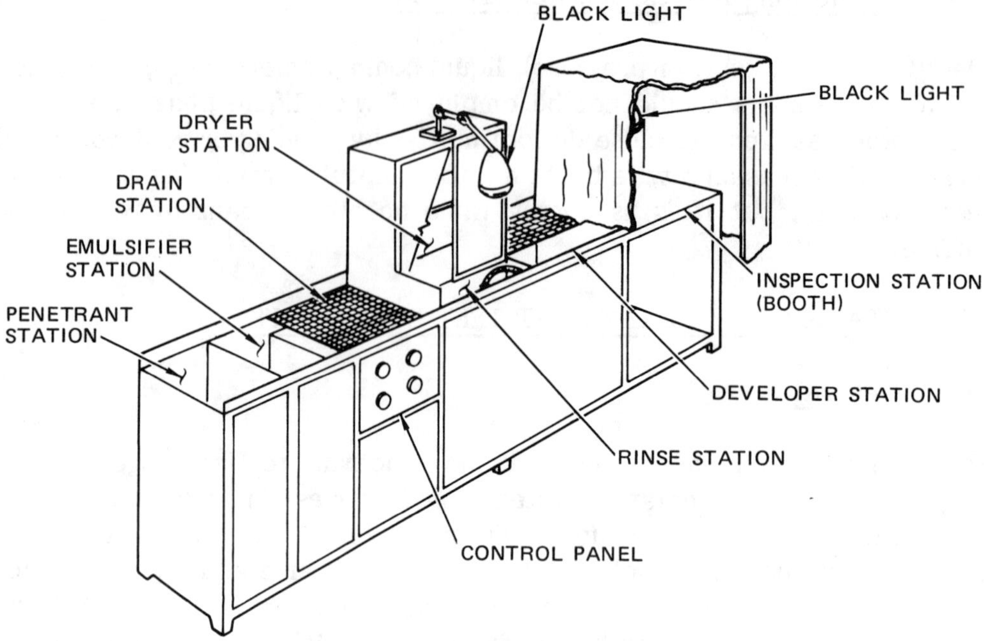

Figure 3-1. Typical Small-Sized Test Equipment Employing Fluorescent Post-Emulsified Penetrant and Dry Developer

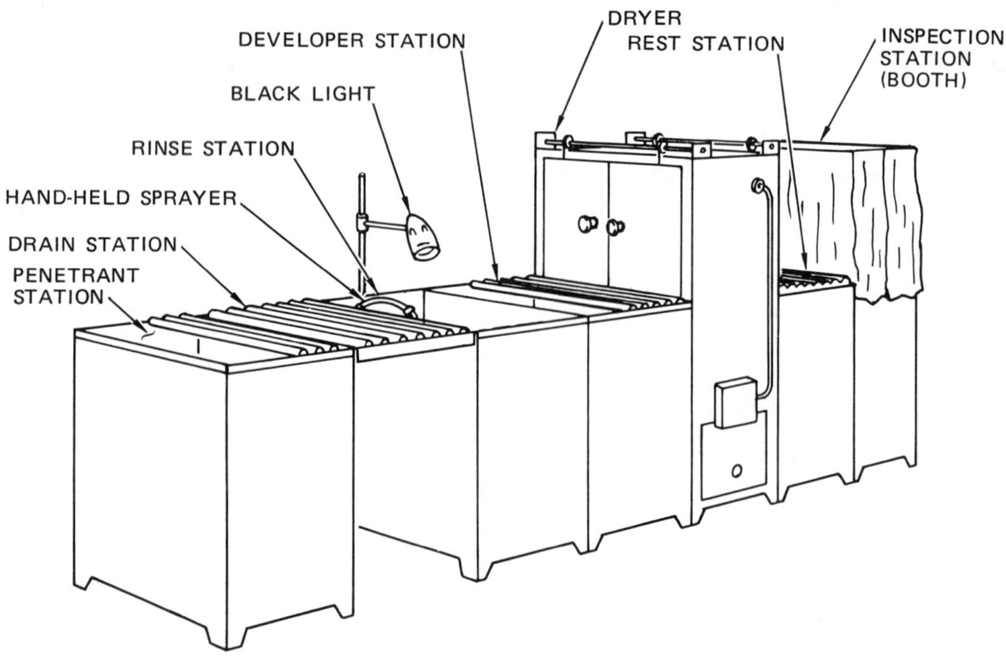

Figure 3-2. Typical Medium-Sized Test Equipment Employing Fluorescent Water-Washable Penetrant and Wet Developer

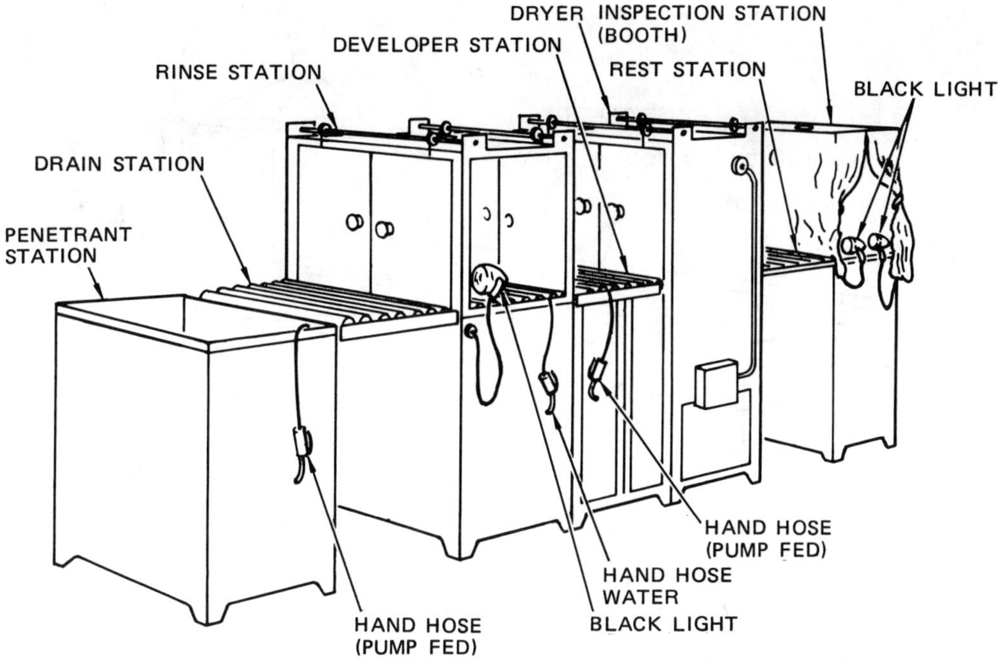

Figure 3-3. Typical Large-Sized Test Equipment Employing Fluorescent Water-Washable Penetrant and Wet Developer

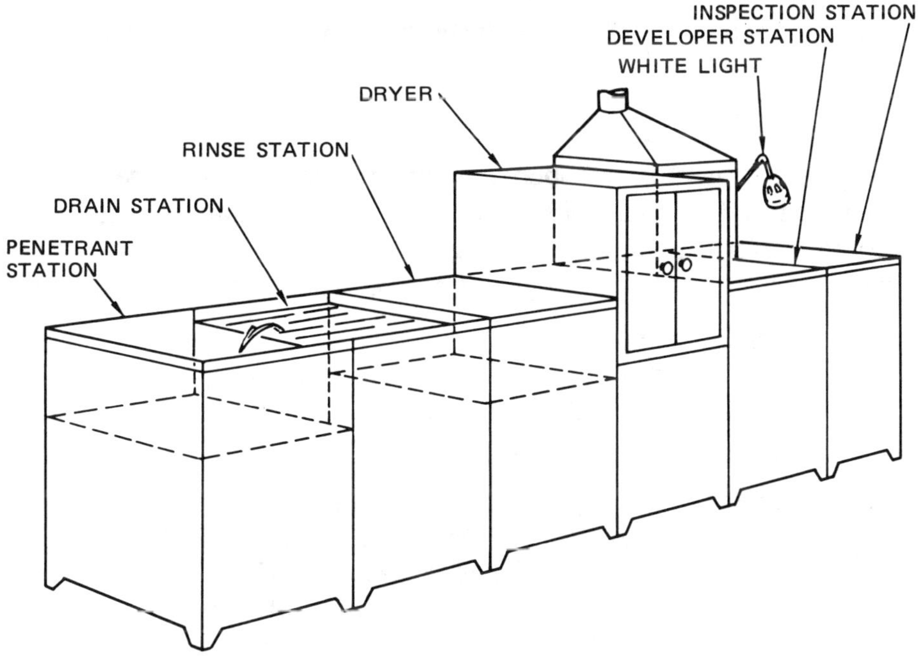

Figure 3-4. Typical Medium-Sized Test Equipment Employing Visible Dye Water-Washable Penetrant and Dry Developer

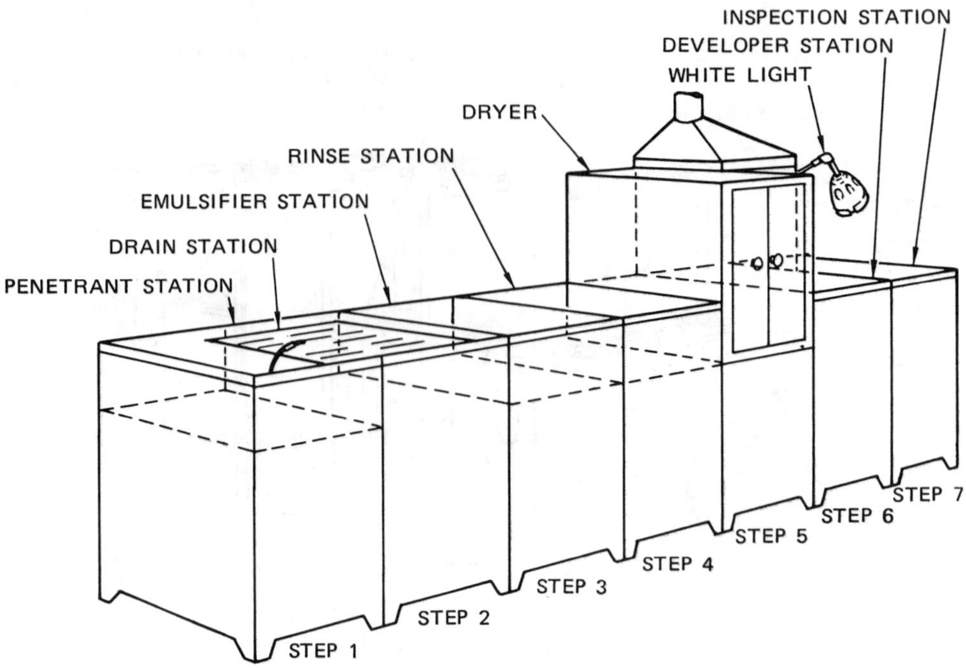

Figure 3-5. Typical Medium-Sized Test Equipment Employing Visible Dye Post-Emulsified Penetrant and Dry Developer

h. Inspection Station (enclosed booth or table with lighting facilities).

i. Postcleaning Station (usually remote from penetrant test station).

3. Auxiliary Equipment

For the purpose of this handbook, auxiliary equipment is defined as the equipment located at penetrant test stations (other than cleaning stations) required to perform penetrant testing. The auxiliary equipment discussed may in some instances be "built-in" at one or more of the test stations.

a. Pumps. Various pumps installed at the penetrant, emulsifier, rinse, and developer stations are used to agitate the solutions, to pump drain-off material into the proper tank for reuse, and to power hand-held sprayers and applicators.

b. Sprayers and Applicators. Sprayers and applicators are frequently employed at the penetrant, emulsifier, rinse, and developer stations. They decrease test time by permitting rapid and even application of penetrant materials and water rinse. Both conventional and electrostatic sprayers are used.

c. Lights. White lights as well as black lights are installed as required to ensure adequate and correct lighting at all stations. When fluorescent materials are used, black light is installed at both the rinse and inspection stations.

d. Timers. One or more 60-minute timers with alarm are used to control penetrant, emulsifier, developing, and drying cycles.

e. Thermostats and Thermometers. These items are required and used to control the temperature of the drying oven and penetrant materials.

f. Exhaust Fans. Exhaust fans are used when testing is performed in closed areas. The fans facilitate removal of fumes and dust.

g. Hydrometers. The hydrometers used in liquid penetrant testing are floating type instruments. (See Figure 3-6.) They are used to measure the specific gravity of water-based wet developers.

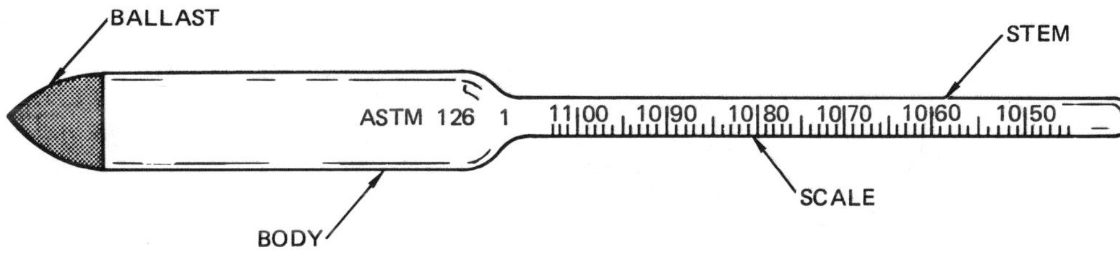

Figure 3-6. Typical Hydrometer

303 PORTABLE PENETRANT TEST EQUIPMENT

1. General

It is possible to perform penetrant tests on a limited basis without stationary equipment. When testing is required at a location remote from stationary equipment, or when only a small portion of a large specimen requires test, portable liquid penetrant kits are used. Both fluorescent and visible dye penetrants are available in kits. The penetrant materials are usually dispensed from pressurized spray cans or applied by brush.

2. <u>Visible Dye Penetrant Kit</u>

The visible dye penetrant test kit is light in weight and contains all the materials necessary for test. (See Figure 3-7.) It consists of a metal box with at least the following:

a. Solvent cleaner or penetrant remover.

b. Visible dye penetrant.

c. Nonaqueous wet developer.

d. Wiping cloths and brushes.

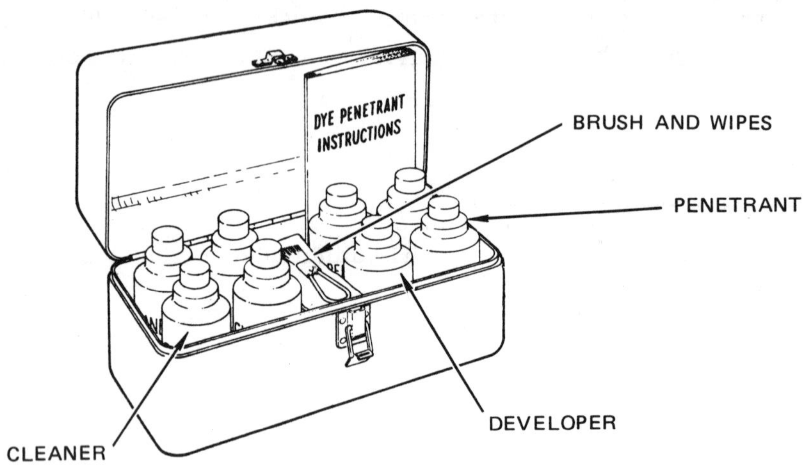

Figure 3-7. Typical Visible Dye Portable Kit

3. <u>Fluorescent Penetrant Kit</u>

The fluorescent penetrant kit combines portability with the high "see-ability" associated with fluorescent materials. The kit holds all the essential materials required for test, including a black light. (See Figure 3-8.) The fluorescent kit consists of a metal box with at least the following:

a. Portable black light and transformer.

b. Solvent cleaner or penetrant remover.

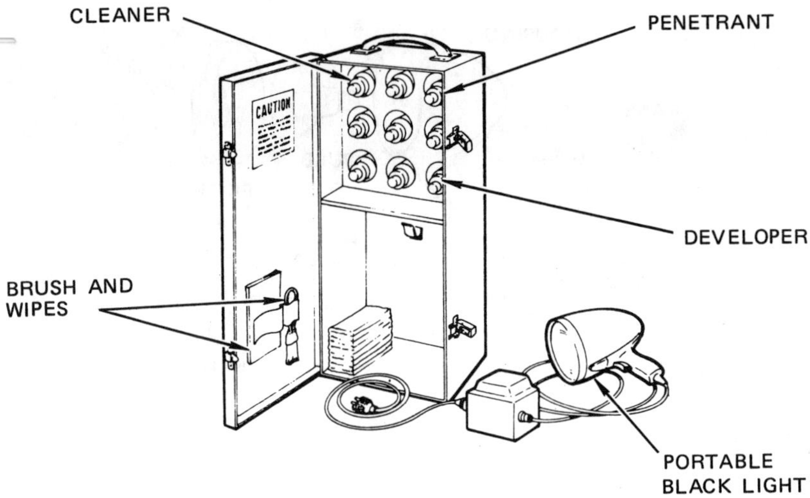

Figure 3-8. Typical Fluorescent Portable Kit

c. Fluorescent penetrant.

d. Nonaqueous wet developer.

e. Dry powder developer.

f. Wiping cloths and brushes.

g. Hood to provide darkened area for viewing indications.

304 BLACK LIGHT

Black light equipment is required in fluorescent penetrant testing, since it supplies light of the correct wavelengths to cause fluorescent materials to fluoresce. The equipment usually consists of a current regulating transformer, a mercury arc bulb, and a filter (see Figure 3-9). The transformer is housed separately and the bulb and filter are contained in a reflector lamp unit. For correct test results the lamp should produce an intensity of at least 800 microwatts per square centimeter at the test surface. The deep red-purple filter is designed to pass only those wavelengths of light that will activate the fluorescent material. It also filters out harmful ultraviolet radiation. Since dust, dirt, and oil greatly reduce the intensity of the emitted light, the filter should be frequently cleaned. In use, the full intensity of the lamp is not attained until the mercury arc is sufficiently heated. At least 5 minutes warmup is required to reach the required arc temperature. Since switching the lamp on and off shortens bulb life, once turned on the lamp is usually left on during the entire test or work period. If the black light is switched off, it may take up to 10 minutes for the bulb to cool sufficiently to reestablish an arc.

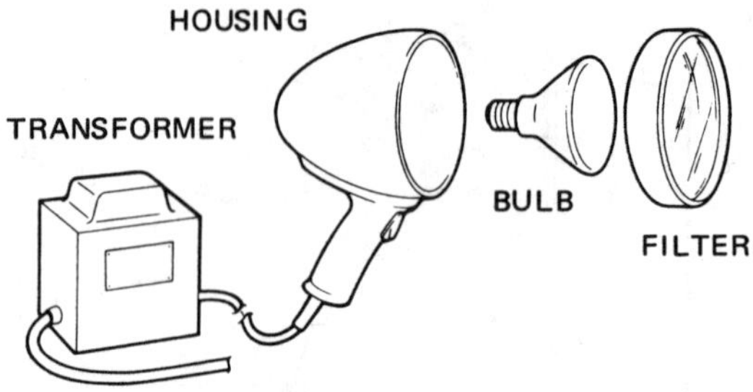

Figure 3-9. Typical Portable Black Light

305 MATERIALS

1. General

The materials used in liquid penetrant testing include penetrants, emulsifiers, removers or cleaners, and developers. They are furnished in either liquid or powder form. The powders — except those used in the dry state — are mixed with a suitable liquid (usually water) prior to use. Most of the materials are available in pressurized spray cans as well as in bulk quantities. Concentrations, usage, and maintenance are in accordance with the manufacturer's directions. Figure 3-10 illustrates the different material combinations and usages.

2. Precleaning and Postcleaning Materials

Except for LOX compatibility, and the chlorine-free requirement in the precleaning and postcleaning of nickel alloys, certain stainless steels, and titanium, no special cleaning materials are required with liquid penetrant testing.

3. Water-Washable Penetrants

Water-washable penetrants are highly penetrating oily liquids containing an emulsifying agent that renders the oily vehicle emulsifiable in water. The simplest to use but least sensitive of these penetrants are the visible dye or color contrast penetrants. They contain a dye, usually a bright red but sometimes a special color such as blue, that can be seen under ordinary white (visible) light. Greatest "seeability" is obtained with fluorescent penetrants that are viewed under black light. The color of fluorescence is usually a brilliant yellowish green. For special applications, there are fluorescent penetrants that glow red or blue. The dual sensitivity penetrants contain a combination of visible and fluorescent dyes. The visible color is usually a bright red and the fluorescent color a yellow to orange-red. They permit gross discontinuities to be detected under visible light, and questionable indications to be resolved under black light.

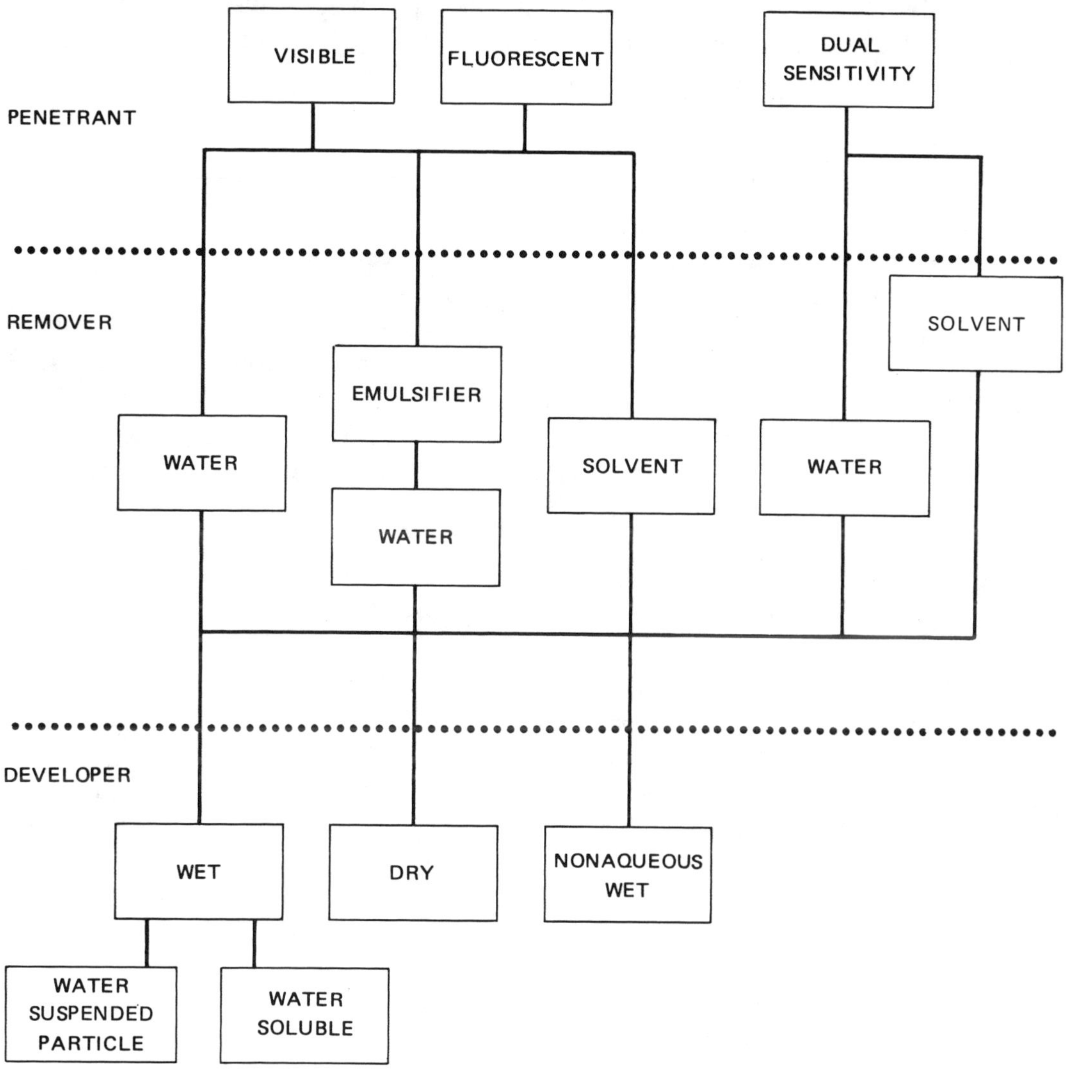

Figure 3-10. Penetrant Material Combinations and Usages

4. Post-Emulsification (Solvent Removable) Penetrants

Post-emulsification penetrants have similar formulations to those of water-washable penetrants except they do not contain the emulsifying agent and consequently are not soluble in water. These penetrants must be treated with a separate emulsifier before they can be removed by a water rinse or wash. Or they can be removed using an approved solvent remover or cleaner. Post-emulsification penetrants are available as either visible dye or fluorescent penetrants.

5. Emulsifiers

Emulsifiers when applied to a post-emulsification penetrant combine with the penetrant so as to make the resultant mixture water washable. The emulsifier, usually dyed orange to contrast with the penetrant, may be either lipophilic — an oil base, or hydrophilic — a detergent water base. The oil-based emulsifiers are usually employed as "contact" emulsifiers, i.e., they begin emulsifying on contact with the penetrant. Emulsification stops when water is applied. The hydrophilic or water-based emulsifiers also can be used as contact emulsifiers; but more often, the emulsifier is diluted with water and sprayed under pressure. Hence the term "hydrophilic scrubber."

6. Solvent Removers (Cleaners)

Solvent removers or cleaners are used in conjunction with post-emulsification penetrants to remove excess penetrant from test article surfaces. Example solvent removers include methylene chloride, isopropyl alcohol, naphtha, mineral spirits (paint thinner) in addition to special-formula, proprietary removers. In selecting a solvent remover, only those materials approved by the penetrant manufacturer can be used.

7. Dry Developer

Dry developer is a fluffy chalk-like powder that is applied to dry test surfaces (after the removal of excess penetrant) for the purpose of absorbing penetrant from discontinuities and enhancing the resultant penetrant indications. Of the different developers available, dry developer is the most adaptable to rough surfaces and automatic processing. It's also the easiest to remove. Sensitivity is about the same as that of the water-soluble developer described in the following paragraph.

8. Water-Based Wet Developers

Water-based wet developers function similarly to dry developer except they are applied prior to drying the test specimen. Two types of developer are available. In one, the developer particles are held in suspension in water and require continuous agitation to keep the particles in suspension. In the other, the developer powder is dissolved in water, forming a solution; once mixed they remain mixed. Of the two water-based wet developers, the water-soluble developer is the more sensitive.

9. Nonaqueous Wet Developer

Nonaqueous wet developer is a suspension of developer particles in a rapid-drying solvent. It is most often employed with solvent-removed processing, and like dry developer, is applied only to dry surfaces. Of all the developers, the nonaqueous wet developer is the most sensitive in detecting fine discontinuities. The evaporation of the solvent carrier helps to draw the penetrant from discontinuities.

10. Special-Purpose Penetrant Materials

In addition to the conventional penetrants, emulsifiers, removers, and developers employed in liquid penetrant testing there are low sulfur and chlorine materials for testing nickel alloys, certain stainless steels, and titanium. Special-purpose inert materials are available for testing articles that come in contact with liquid oxygen, rubber, or plastic. Food compatible materials are also available. There are high temperature penetrants for testing hot welds, etc., and special penetrants for testing at low temperatures. There are supersensitive penetrants for detecting extremely fine discontinuities, and penetrants that provide sufficient contrast and sensitivity without a developer. There are low-energy emulsifiers and inhibited-solvent removers to slow down emulsification and the removal of excess penetrant. There are also wax and plastic film developers that absorb and fix penetrant indications to provide permanent records. The selection and usage of these materials is largely dependent on the particular process used and the controlling specifications or standards.

CHAPTER 4: TECHNIQUES

TABLE OF CONTENTS

LIST OF FIGURES

LIST OF TABLES

CHAPTER 4: TECHNIQUES

400 GENERAL

The techniques discussed in this chapter are based on typical liquid penetrant testing procedures used throughout industry. Included are techniques involving the use of visible dye, fluorescent, and dual sensitivity penetrants; and water-washable, post-emulsified, and solvent-removed processing. Also included are discussions on the fixing and recording of indications.

401 SURFACE PREPARATION

1. General

The effectiveness of liquid penetrant testing is based upon the ability of the penetrant to enter surface discontinuities. The article to be tested must be clean and free from foreign matter. All paint, carbon, oil, varnish, oxide, plating, water, dirt, and similar coatings must be removed prior to the application of penetrant. The cleaning technique used is, in each case, determined by the composition of the article under test and the type of soil to be removed. Any cleaning process that leaves the surface of the article clean and dry, that does not harm the article, and that does not use materials that are incompatible with the penetrant materials, is acceptable. Following the test, postcleaning is employed to remove the residue of penetrant materials. Postcleaning is particularly important when test articles are destined for use in an oxygen environment. Though many specimens will receive further processing, such as etching or special cleaning prior to use, the cleanliness of any specimen after completion of a penetrant test is the responsibility of test personnel.

2. Detergent Cleaning

Detergent cleaning may be used to clean almost any specimen. Since the cleaners may be either acid or alkaline in nature, however, precautions must be taken to ensure that the selected detergent is noncorrosive to the specimen being cleaned. Detergent cleaning is most effective when it is a hot process accomplished in a washing machine, though it may also be used with scrub, rinse, and wipe techniques. After detergent cleaning, the specimen is carefully rinsed and dried. The drying process should be of sufficient time duration that all moisture is driven from the discontinuities.

3. Vapor Degreasing

Vapor degreasing is also an effective means of precleaning. The process not only thoroughly cleans, it heats the article so that after cleaning no moisture remains in discontinuities. Vapor degreasing is the preferred method for removing organic soils such as oil and grease and should be used whenever practicable. The only precaution required in the use of the process is that caused by the need of using only those degreasing materials that are not harmful to the specimen being cleaned.

4. Steam Cleaning

Steam cleaning is an excellent method of cleaning usually employed to clean large articles, or portions of large articles, that cannot conveniently be vapor degreased or washed with detergents. Routine steam cleaning procedures usually suffice for penetrant precleaning. As with any cleaning process involving water, the specimen must be thoroughly dried after the cleaning process is completed.

5. Ultrasonic Cleaning

Ultrasonic cleaning is often combined with a solvent or detergent bath to improve cleaning efficiency and reduce cleaning time. The method works best with water and detergent cleaning when contaminants to be removed are inorganic, and with solvents when contaminants are organic. Following cleaning, it is recommended that test articles be heated to aid the evaporation of cleaning fluids.

6. Rust and Surface Scale Removal

Rust removers (descaling solutions, either alkaline or acid), pickling solutions (acid), and sometimes wire brushing are used to remove rust and surface scale. Wire brushing is accomplished with a minimum of pressure to avoid closing surface discontinuities or filling them with smeared metal. Descaling solutions are chosen so that they are noncorrosive to the article being cleaned. Regardless of the method selected for rust and scale removal, after the process is completed the specimen must be clean, dry, and so treated that surface discontinuities are not clogged, filled, or contaminated.

7. Paint Removal

Any method of paint removal that does not harm the test article is satisfactory. Chemical means such as solvent stripping and dissolving type hot-tank stripping are preferred since any mechanical removal process may adversely affect the surface of the specimen.

8. Etching

Etching is normally required on soft metallic materials (such as aluminum and magnesium) and materials that tend to smear (such as titanium), and which have been mechanically processed by machining, grinding, or similar procedure. The etching is accomplished with either an acid or an alkaline solution, which is then neutralized. After neutralization, the article must be water washed and dried, or otherwise cleaned, to remove all traces of the etching and neutralizing agents.

402 APPLICATION OF PENETRANTS

1. General

Penetrants are applied by spraying, swabbing, brushing, or dipping (immersion). The area under test is covered with penetrant and the penetrant is allowed to remain for a predetermined amount of time called "dwell time." The means of application and the length of dwell are determined by the test article, the type discontinuities to be detected, the penetrant used, and temperature. The terminology used in penetrant application is listed in Table 4-1.

Table 4-1. Liquid Penetrant Application Terminology

Technique	Soak Time	Dwell Time	Penetration Time
Immersion	Period Article is in Bath	Draining	Soak Time Plus Dwell Time
All others		Time Penetrant Remains on Article	Dwell Time

2. Spraying

Spraying of penetrant when accomplished at the penetrant tank of stationary equipment refers to the use of a hose and nozzle through which penetrant is circulated by a low pressure pump — usually the same pump that agitates the penetrant solution in the tank. The penetrant is flowed on the specimen so that all of the test area is covered. No particular precautions except those of cleanliness and neatness need be observed in this flow-on process. Spraying also is used to define the application of penetrant from pressurized spray cans. Again the penetrant is applied so that all of the test area is covered, but personnel must make allowances for the pressure remaining in the can and the distance the can is held from the specimen. Usually, pressurized spray cans are used in areas where fans or blowers remove fumes, or in open areas where spot testing (testing a small area of a large specimen) is taking place.

3. Swabbing or Brushing

Penetrants may be applied by swabbing with rags or cotton waste, or by brushing. Either method is acceptable when spray or dip equipment is not available. Usually, swabbing or brushing is used when testing a small, specific area of the specimen.

4. Immersion

The best procedure for applying penetrant is to immerse the test article or specimen into a tank of penetrant. Small specimens are placed in an open wire basket for dipping; large specimens are handled by hand or, if required, by cranes and suitable clamping devices. This method is impractical when dealing with large articles or assemblies, and is wasteful when only small areas of a large specimen are to be tested. It is, however, the most thorough, and certain, means of applying penetrant and is used whenever possible.

5. Penetration (Dwell) Time

The period of time during which the penetrant is permitted to remain on the specimen is a vital part of the test. This time, known as dwell time, is directly related to the size and shape of the discontinuities anticipated, since the dimensions of the discontinuities determine the rapidity with which penetration occurs. Tight cracklike discontinuities may require in excess of 30 minutes for penetration to an extent that an adequate indication can be expected. Gross discontinuities may be suitably penetrated in 3 to 5 minutes. Dwell time in each instance is determined by the anticipated discontinuities and the penetrant manufacturer's recomendations. Typical minimum penetration times are shown in Table 4-2.

a. Heating the test specimen accelerates penetration and shortens dwell time. The practice, however, is generally not recommended since heating may cause evaporation of penetrant and thereby reduce sensitivity.

b. Ambient temperature and humidity also affect penetration time. Generally, the higher the ambient temperature, the shorter the dwell time required. Too high a temperature or too low a humidity, however, causes the penetrant to dry too rapidly and testing becomes difficult if not impossible. For liquid penetrant testing to be reliable, the penetrant must remain wet. This sometimes requires the rewetting of test surfaces. If the penetrant has been allowed to dry, the test must be started over beginning with surface preparation.

403 APPLICATION OF EMULSIFIER

When post-emulsification penetrants are used, an emulsifier must be applied to the penetrant to make it removable by water rinse. Emulsifier is applied by any of the means used to apply penetrant except brushing but dipping is preferred. The amount of time that the emulsifier is permitted to remain (dwell) prior to the removal process is usually in the range of 1 to 3 minutes, and seldom exceeds 5 minutes. The exact emulsification dwell time must be determined by experiment.

404 REMOVAL OF PENETRANTS

1. General

Following application of the penetrant and elapse of sufficient time for penetration, the penetrant is removed from the surface of the specimen. This operation is meant to remove the penetrant from the surface without disturbing any penetrant that has entered a discontinuity. Complete removal of the surface penetrant is effected to ensure against formation of nonrelevant indications.

2. Water-Washable Process

The penetrants employed in the water-washable process have their own built-in emulsifier. The penetrant is soluble in water and removal is usually accomplished by a water rinse. Care is taken in applying the rinse to

Table 4-2. Typical Minimum Penetration Times

Material	Form	Type of Discontinuity	Water-Washable Penetration Time*	Post-Emulsified Penetration Time*	Solvent-Removed Penetration Time*
Aluminum	Castings	Porosity Cold Shuts	5 to 15 min 5 to 15	5 Min ** 5 **	3 Min. 3
	Extrustions & Forgings	Laps	NR***	10	7
	Welds	Lack of Fusion Porosity	30 30	5 5	3 3
	All	Cracks Fatigue Cracks	30 NR***	10 30	5 5
Magnesium	Castings	Porosity Cold Shuts	15 15	5** 5**	3 3
	Extrusions & Forgings	Laps	NR***	10	7
	Welds	Lack of Fusion Porosity	30 30	10 10	5 5
	All	Cracks Fatigue Cracks	30 NR***	10 30	5 7
Steel	Castings	Porosity Cold Shuts	30 30	10** 10**	5 7
	Extrusions & Forgings	Laps	NR***	10	7
	Welds	Lack of Fusion Porosity	60 60	20 20	7 7
	All	Cracks Fatigue Cracks	30 NR***	20 30	7 10
Brass & Bronze	Castings	Porosity Cold Shuts	10 10	5** 5**	3 3
	Extrusions & Forgings	Laps	NR***	10	7
	Brazed Parts	Lack of Fusion Porosity	15 15	10 10	3 3
	All	Cracks	30	10	3
Plastics	All	Cracks	5 to 30	5	5
Glass	All	Cracks	5 to 30	5	5
Carbide-tipped Tools		Lack of Fusion Porosity	30 30	5 5	3 3
		Cracks	30	20	5
Titanium & High Temp Alloys	All		NR***	20 to 30	15
All Metals	All	Stress or Inter-Granular Corrosion	NR***	240	240

*For Parts Having a Temperature of 60 to 125°F (16 to 25°C)
**Precision Castings Only
***NR – Not Recommended

ensure that the spray volume and force does not wash the penetrant from discontinuities. Thirty to fifty pounds per square inch maximum pressure (205 to 345 kPa) is considered a safe pressure for the water rinse. The rinse is applied through the use of an adjustable spray nozzle held so that the spray reaches the surface plane of the specimen at an angle of 45 degrees.

3. Post-Emulsified Process

The penetrants employed in the post-emulsified process do not contain an emulsifying agent. The penetrant is not soluble in water. Removal is in most instances a two-step process. The emulsifier, usually lipophilic (an oil base), is applied as described in paragraph 403 and, after suitable dwell time, the resultant penetrant-emulsifier mixture is removed by water rinse as described in paragraph 404.2. Sometimes a hydrophilic (water base) emulsifier is diluted to the point that simple contact with penetrant does not make the penetrant water washable. Application must be accompanied by some form of mechanical agitation or scrubbing. Hence the term "hydrophilic scrubber." Usually, the emulsifier is added to the water rinse and sprayed under pressure. By controlling solution strength and the duration of spray, the amount of penetrant removed is controlled.

4. Solvent-Removed Process

Post-emulsification type penetrants are also employed in the solvent-removed process. The penetrant remover is a solvent designated by the penetrant manufacturer. Prior to the use of the solvent, excess penetrant is wiped off, the specimen is then cleaned with clean, lint-free towels dampened with solvent. The solvent is never applied directly to the specimen since it might wash out or dilute the penetrant in a discontinuity.

5. Visual Inspection

Excess surface penetrant can result in the formation of nonrelevant indications that could obscure or hide true discontinuity indications. When fluorescent penetrants are used, it is necessary to observe the specimen under black light during the penetrant removal operation to ensure complete removal of excess penetrant. For visible dye penetrants, the absence of penetrant (red) traces on the wiping materials ensures complete penetrant removal.

405 APPLICATION OF DEVELOPER

1. General

As mentioned in previous chapters, some penetrants provide sufficient discontinuity indications without a developer. They are self-developing. But generally, when maximum sensitivity is desired, a developer is required. The developer assists in the detection of penetrant retained in discontinuities by aiding in the capillary bleed-out process (the developer acts as a blotting agent), and by accentuating the presence of penetrant in a discontinuity. Developer accentuates the presence of a discontinuity because it causes the penetrant from the discontinuity to spread out over a greater area. It also serves as a color contrast background for the visible dye used in the visible dye processes and for the fluorescent material used in the fluorescent processes. Developer is available in both dry and liquid forms and the selection of developer is in accordance with the manufacturer's recommendation for the type penetrant used. When a dry or nonaqueous wet developer is used, the specimen must be completely dry before the developer is applied. When a water-based wet developer is used, it is applied immediately after penetrant removal is accomplished and prior to the drying operation.

2. Dry Developer

Dry developer, being a loose, fluffy talcose powder with high absorbent properties, is applied to a specimen by dusting, blowing, or dipping the specimen. The application is usually accomplished in a booth with a blower or fan arrangement that removes loose powder from the atmosphere. No preparation of the powder is necessary and the only requirement is that it be evenly distributed over the test surface, which must be completely dry.

3. Nonaqueous Wet Developer

Nonaqueous wet developer is a suspension of absorptive white powder in a solvent vehicle. It is usually applied by spraying from a pressurized spray can or other spraying device such as a paint spray gun. When used in bulk form, care must be exercised to keep the powder thoroughly mixed in the solvent. The developer is applied so as to form a thin white coating on the specimen without soaking the test surface. When properly mixed and applied, nonaqueous wet developer is the most sensitive of all the developers in detecting fine discontinuities.

4. Water-Based Wet Developer

Water-based wet developer may be either a suspension of absorptive white powder in water, or a water-soluble absorptive white powder mixed with water. The suspension type requires mild agitation prior to and during use to keep the powder particles in suspension; the water-soluble developer does not. The water-soluble powder, once mixed with the water, remains in solution. After excess penetrant is removed from the specimen, and while it is still wet, wet developer is applied by either dip (immersion), flow-on, or spray techniques. These fast and effective methods of application, combined with the time saved by applying developer to the wet specimen, make water-based wet developer well suited for use in rapid, production line testing. Wet developer is applied so as to form a smooth, even coating, and particular care is taken to avoid concentrations of developer in dished or hollowed areas of the specimen. Such concentrations of developer mask penetrant indications and are to be avoided.

406 DRYING

When dry or nonaqueous wet developer is used, the specimen is dried after removal of excess penetrant and prior to application of the developer. When water-based wet developer is used, the specimen is dried after the developer has been applied. Any means of drying that does not interfere with the test process by overheating, or by contamination of materials, is acceptable, but controlled drying at even regulated temperatures is preferred. A thermostat controlled dryer with a temperature range up to 225°F (107°C) is usually employed in stationary test installations. Required drying time is determined by the size and shape of the specimen, and by the nature of its suspected discontinuities. It should be of sufficient duration to dry the surface of the specimen without affecting the penetrant in the discontinuities.

407 PENETRANT TESTING PROCESSES

1. General

The different processes employed in liquid penetrant testing are identified by the method of penetrant removal used (water-washable, post-emulsified, or solvent-removed) and the type of dye (visible dye (color contrast), fluorescent, or dual sensitivity). The basic steps involved are illustrated in Figures 4-1 and 4-2 while step-by-step procedures are contained in the following paragraphs. Table 4-3 lists the preferred processes for various penetrant test problems.

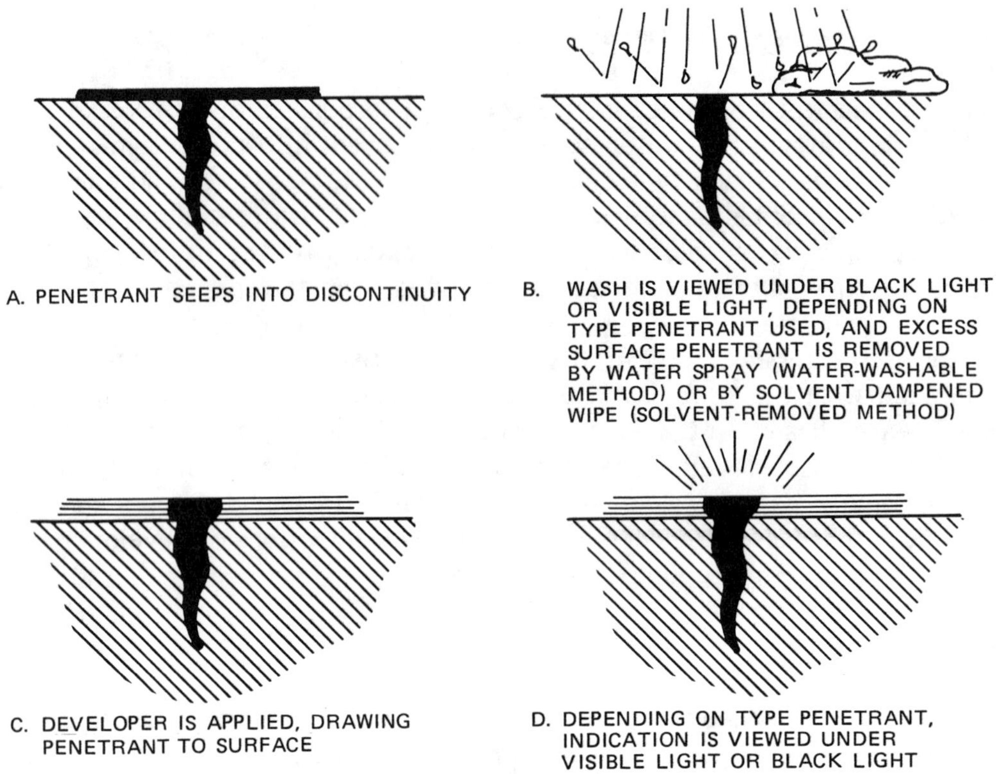

Figure 4-1. Water-Washable and Solvent-Removed Processes

2. Water-Washable Fluorescent Penetrant Test

The characteristic advantages and disadvantages of water-washable fluorescent penetrant tests are listed in Table 4-4.

a. Penetrant Application. Either immersion, flow-on, spray, or brushing technique is used to apply the penetrant to the precleaned, dry specimen. The penetrant is applied evenly over the entire test area.

b. Dwell Time. The penetrant is left on the specimen for the required length of dwell time. A broad guide to correct dwell time is contained in Table 4-2 but the specimen size, composition, and discontinuities, and the temperature of the specimen and the test area all affect required dwell time.

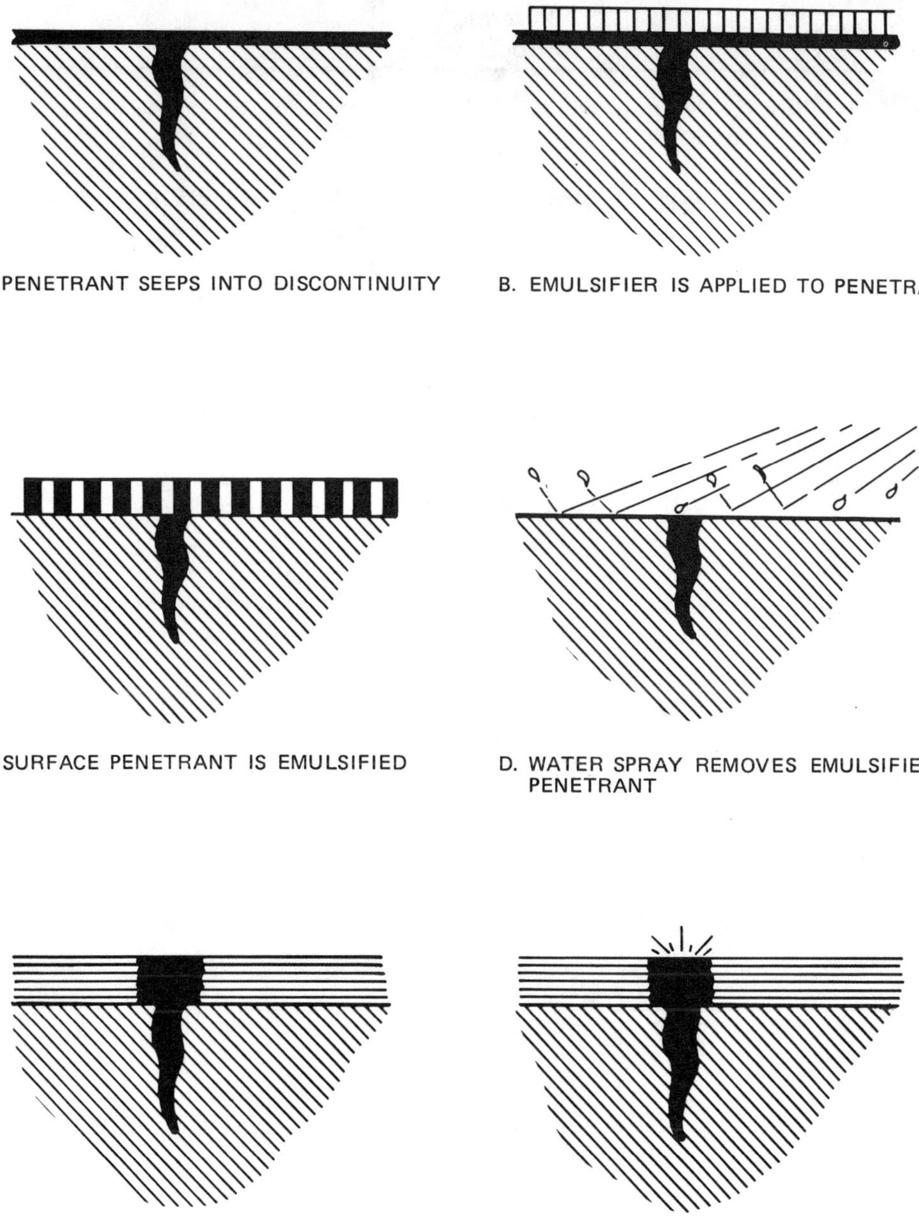

A. PENETRANT SEEPS INTO DISCONTINUITY

B. EMULSIFIER IS APPLIED TO PENETRANT

C. SURFACE PENETRANT IS EMULSIFIED

D. WATER SPRAY REMOVES EMULSIFIED PENETRANT

E. DEVELOPER DRAWS PENETRANT FROM DISCONTINUITY

F. DEPENDING ON TYPE PENETRANT, INDICATION IS VIEWED UNDER VISIBLE LIGHT OR BLACK LIGHT

Figure 4-2. Post-Emulsified Process

Table 4-3. Process Selection Guide

Testing Problem	Preferred Process	Remarks
High Production of Many Small Articles Required	Water Washable	Small Articles Handled in Baskets
High Production of Large Individual Articles	Post Emulsified	Large Forgings, Extrusions, etc.
Highest Sensitivity to Fine Discontinuities	Post Emulsified	Brightest Indication Most Sensitive
Shallow Discontinuities Scratches, etc., Must be Detected	Post Emulsified	Depth of Emulsification Can be Controlled
Articles Having a Rough Surface	Water Washable	
Articles Having Threads and Keyways	Water Washable	Post Emulsified Penetrant Might Lodge in Corners
Articles Having Medium Rough Surfaces	Water Washable or Post Emulsified	Choice Depends Upon Production and Sensitivity Requirements
Spot Testing of Local Areas Desired	Solvent Removed	
Portable Equipment Necessary	Solvent Removed	
Water and Electricity not Available	Solvent Removed	
Anodized Articles, Cracked after Anodizing, to be Tested	1. Solvent Removed 2. Post Emulsified 3. Water Washable	Order of Preference Indicated
Repeated Application of Process is Desired	Solvent Removed	Five or Six Repeats Should be the Limit
Leak Detection	Water Washable or Post Emulsified	

Table 4-4. Characteristics of Water-Washable Fluorescent Penetrant Tests

Advantages	Disadvantages
Fluorescence Ensures Visibility	Requires Darkened Area for Inspection
Easily Washed with Water	Not Reliable for Detecting Scratches and Similar Shallow Surface Discontinuities
Good for Volume Testing of Small Specimens	Not reliable on Reruns of Specimens
Good on Rough Surfaces	Not Reliable on Anodized Surfaces
Good on Keyways and Threads	Acids and Chromates Affect Sensitivity
Good on Wide Range of Discontinuities	Easily Over-washed
Fast, Single Step Process	Penetrant Subject to Water Contamination
Relatively Inexpensive	
Available in Lox Compatible Form	

c. Penetrant Removal. Excess penetrant (all penetrant except that in discontinuities) is washed from the specimen after dwell time has elapsed. Water at 60 to 110°F (16 to 43°C) and a pressure not exceeding 50 psi (345 kPa) is applied from a spray nozzle. The nozzle is held so that the water strikes the surface of the specimen at an angle of approximately 45 degrees. Care is taken to avoid over-washing, which causes washout of penetrant from discontinuities. The wash process is accomplished under black light so that the operator can observe when the excess penetrant is completely removed.

d. Drying. Upon completion of the wash process the specimen is dried prior to the application of either dry or nonaqueous wet developer. If water-based wet developer is used, it is applied to the still damp specimen immediately after the penetrant removal wash. Drying is best accomplished in a thermostat-controlled oven at a temperature between 150 and 225°F (66 to 107°C). Drying time is determined by the size and composition of the specimen, and visual observation usually fixes the length of the drying cycle. Excessive heat or too long a drying time tends to bake the penetrant out of discontinuities.

e. Developer Application. When the drying process is complete the specimen is ready for the application of either dry or nonaqueous wet developer. When water-based wet developer is used, it is applied to the wet specimen immediately after excess penetrant is removed.

(1) Dry developer is applied to the specimen by brushing with a soft brush, by use of a powder gun, or by dipping the specimen in a tank of the developer, and removing excess powder with a low pressure air flow.

(2) Nonaqueous wet developer is applied by spraying. It is applied sparingly so that a thin coating covers all of the specimen test area. When using nonaqueous wet developer the specimen is to be cool enough to prevent too rapid evaporation of the developer vehicle.

(3) Water-based wet developer is applied to the specimen as it comes from the wash cycle, either by immersion or flow-on. The developer is applied so as to form a smooth even coating over the entire test area. After the developer is applied, the specimen is dried as described in paragraph 407.2.d.

f. Inspection. After sufficient time has passed for developer action to bring the penetrant from discontinuities as indications, the specimen is ready for inspection under black light. The interpretation of various indications discovered during inspection is discussed in Chapter 5. The efficiency of the inspection operation is controlled by the variables of the human eye. These variables are further complicated by the average person's lack of understanding of eye fatigue and of the time required for the iris of the eye to dilate to a point of maximum vision in the darkness of the black light inspection booth. For maximum visual efficiency the operator must:

(1) Let eyes become accustomed to the darkness by entering the darkened area (booth) at least 5 minutes prior to examining the specimen under the black light.

(2) Avoid looking directly into the black light source since the eyeball contains a fluid that fluoresces if black light shines directly into the eye.

3. Post-Emulsified Fluorescent Penetrant Test

The characteristic advantages and disadvantages of post-emulsified fluorescent penetrant tests are listed in Table 4-5. This process is identical with that of the water-washable fluorescent penetrant test except for the inclusion of an emulsification step after the completion of penetrant dwell time and before penetrant removal.

Table 4-5. Characteristics of Post-Emulsified Fluorescent Penetrant Tests

Advantages	Disadvantages
Fluorescence Ensures Visibility High Sensitivity for Very Fine Discontinuities Good on Wide Shallow Discontinuities Easily Washed with Water After Emulsification Short Penetration Time Cannot be Easily Over-washed	Requires Darkened Area for Inspection Two Step Process Equipment Required for Emulsifier Application Difficult to Remove Penetrant from Threads, Keyways, Blind Holes and Rough Surfaces

a. Penetrant Application. See paragraph 407.2.a.

b. Dwell Time. See paragraph 407.2.b.

c. Emulsifier Application. After the elapse of sufficient dwell time, emulsifier is applied to the penetrant coated specimen. Immersion, flow-on, or spray technique is used to apply the emulsifier in an even coating. The particular technique employed is determined by the number and size of the specimens under test.

d. Emulsifier Dwell Time. The length of time the emulsifier is left to dwell before commencing the penetrant removal cycle is determined by the emulsifier used and the type discontinuities suspected. Detection of shallow, wide dents, machine marks, and nicks requires a minimum emulsification time. Detection of fine, light cracks requires emulsification time of sufficient duration that superficial discontinuities are washed clean during the penetrant removal, but the time is not to be so long that the penetrant in the cracks is affected. One to 3 minutes emulsification dwell time is usually required, though rough surfaced articles may require 5 minutes or more. Actual time must be determined by experiment.

e. Penetrant Removal. See paragraph 407.2.c.

f. Drying. See paragraph 407.2.d.

g. Developer Application. See paragraph 407.2.e.

h. Inspection. See paragraph 407.2.f.

4. Solvent-Removed Fluorescent Penetrant Test

The characteristic advantages and disadvantages of solvent-removed fluorescent penetrant tests are listed in Table 4-6.

Table 4-6. Characteristics of Solvent-Removed Fluorescent Penetrant Tests

Advantages	Disadvantages
Fluorescence Ensures Visibility	Requires Darkened Area for Inspection
Portability	Flammable Materials
No Water Required	Removal of Excess Surface Penetrant is Time Consuming
Good On Anodized Specimens	Materials Cannot be Used in Open Tanks
Good For Spot Checking	Difficult to use on Rough Surfaces Such as Cast Magnesium
Specimens Can be Rerun	

a. Penetrant Application. Solvent-removed penetrant may be applied by brush-on technique but is more often applied by use of a spray gun or pressurized spray can. With any application process, correct application covers the test surface with an even coat of penetrant. When a spray gun or pressurized can is used, the gun or can is held approximately 12 inches (30 cm) from the specimen and moved slowly from side to side until the specimen is evenly coated.

b. Dwell Time. See paragraph 407.2.b.

c. Penetrant Removal. Excess penetrant is removed from the specimen, after suitable dwell time has elapsed, by wiping with absorbent, lint-free towels. After the bulk of the excess penetrant is wiped off, clean, lint-free towels are moistened with the companion solvent of the penetrant (solvent specified by the penetrant manufacturer) and the specimen is wiped clean. Solvent is never applied directly to the specimen. The removal process is accomplished under black light so the operator can observe that all excess penetrant is removed.

d. Developer Application. Usually only dry or nonaqueous wet developer is used with solvent removed penetrants. A thin coating of developer is either dusted or sprayed on the test area of the specimen.

e. Inspection. See paragraph 407.2.f.

5. Visible Dye Penetrant Tests

The characteristic advantages and disadvantages of visible dye penetrants are the same as those listed in Tables 4-4, 4-5, and 4-6 for their fluorescent counterparts, except that visible dye penetrants are less sensitive, not as brilliantly visible, and do not require the use of black light.

a. Water-Washable Visible Dye Penetrant Test. Procedures for use of water-washable visible dye penetrants are identical with those listed in paragraphs 407.2.a through f, except there is no black light requirement.

b. Post-Emulsified Visible Dye Penetrant Test. Procedures for use of post-emulsified visible dye penetrants are identical with those listed in paragraphs 407.3.a through h, except there is no black light requirement.

c. Solvent Removed Visible Dye Penetrant Test. Procedures for use of solvent removed visible dye penetrants are identical with those listed in paragraphs 407.4.a through e, except that there is no black light requirement.

6. Dual Sensitivity Penetrant Tests

Dual sensitivity penetrant test procedures are identical to those listed in paragraphs 407.2.a through f, except that indications can be viewed under visible light as well as black light. The advantage in using a dual sensitivity penetrant is that it provides two different levels of sensitivity in one test. Gross discontinuities can be located under visible light, and examinations for fine cracks, porosity, and penetrant residue carried out under black light.

408 OXYGEN COMPATIBLE PENETRANTS

There is no basic difference in the techniques used with liquid and gaseous oxygen compatible penetrant materials and those used with other penetrant materials, except for the special precautions required to avoid contamination of the test materials.

409 LEAK-THROUGH TECHNIQUE

Frequently, articles are so designed that the penetrant solution may be poured into them and the outer surface examined for evidence of leak-through. Use of liquid penetrant in this manner detects leaks only, and little or no knowledge concerning other quality characteristics of the specimen is obtained. The leak-through technique is well suited for finding leaks in such articles as tanks, piping, tubing, and hollow castings. Figure 4-3 illustrates the liquid penetrant leak-through test as used on a large plate section.

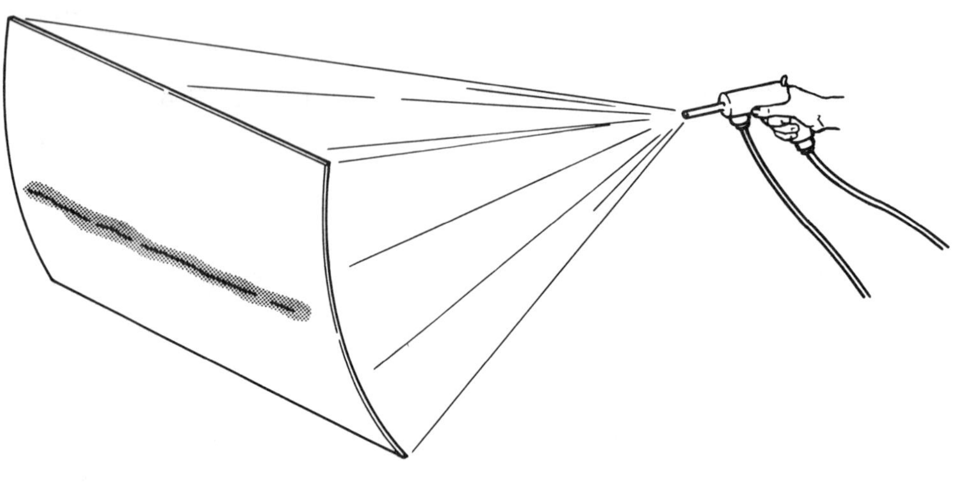

Figure 4-3. Leak-Through Test

410 FIXING AND RECORDING INDICATIONS

Liquid penetrant testing increasingly involves the fixing and recording of indications for use in test reports and inspection records. Toward this end, both black-and-white and color film are used, with the self-developing films providing the ultimate in convenience. In addition to photographs, there are special wax and plastic film developers that absorb and fix penetrant indications to provide permanent records. Indications are sometimes fixed using a "strippable" lacquer. Several coats of lacquer are applied over the indication and allowed to dry. The coating is then lifted, bringing the indication with it. Another method consists of spraying the indication with a "fixer," and when dry, lifting the indication with a piece of transparent tape. The tape, with the indication, is then transferred to the appropriate test document. In the recording of indications, however, it must be remembered that time is of the essence. Delays will cause indications to become blurred and distorted through the continued bleeding of penetrant. As a result, indications will not be representative of the discontinuities from which they came. The fixing and recording of indications should be accomplished as soon as possible following the required development time.

CHAPTER 5: INTERPRETATION OF TEST RESULTS

TABLE OF CONTENTS

LIST OF FIGURES

LIST OF FIGURES (CONT)

CHAPTER 5: INTERPRETATION OF TEST RESULTS

500 GENERAL

The terms "interpretation" and "evaluation" are often confused by testing personnel. Actually, the terms refer to two entirely different steps in the testing process. To interpret an indication means to decide what condition caused it. It may be a crack, porosity, lack of bond, or merely penetrant remaining on the surface due to some nonrelevant cause. Evaluation follows interpretation. If a discontinuity exists, its effect on the usefulness of the article requires evaluation, i.e., the article is either accepted as is, reworked, or scrapped. The success and reliability of the interpretation and evaluation of liquid penetrant test indications depend upon the thoroughness of the process. The liquid penetrant test is not a method by which a specimen is processed through a machine that separates the good article from the bad. Testing personnel are required to carefully process each specimen, interpret indications, evaluate the seriousness of discontinuities, and determine disposition.

501 INDICATIONS

1. General

Since penetrant can only indicate surface discontinuities, an indication is caused by a discontinuity in the surface, or by penetrant remaining on the surface from some nonrelevant cause. Proper interpretation, however, requires familiarity with the manufacturing processes, the types of discontinuities likely to occur, and their appearance. It requires a knowledge of how different materials fail and where. It requires the careful control of development time. The effects of these variables are discussed in the following paragraphs.

2. Development Time

Proper development time is very important in the interpretation of indications. This is the time allowed between the application of developer and the actual viewing of indications. If too little time is allowed, indications may not have time to develop. Too much time will cause indications to become blurred or distorted. The correct time depends on the developer used. With water-based (aqueous) developers, the time in the drying oven is usually sufficient. With a dry or nonaqueous wet developer, the accepted norm is 7 to 30 minutes. (A "rule of thumb" is to use a time equal to one-half the penetration dwell time used.)

3. False Indications

The most common source of false indications is poor washing of water-washable and post-emulsified penetrants. The use of black light during the washing process, when using fluorescent penetrant, is very important. The operator can easily tell whether a good rinse is obtained or whether patches of fluorescence remain on the specimen. With penetrants requiring solvent removal, the removal process is much more likely to be thorough. To guard against confusion resulting from fluorescent or color spots other than true indications, care is taken so that no outside contamination occurs. Typical sources of contamination are:

a. Penetrant on hands of operator.

b. Contamination of wet or dry developer.

c. Penetrant rubbing off of an indication on one specimen to a clean portion of the surface of another specimen.

d. Penetrant spots on the inspection table.

(1) To avoid contamination, its causes are eliminated or guarded against. Process tanks and inspection areas are kept clean; only lint-free wiping cloths or rags are used; and specimens are kept free of fingerprints and tool marks.

(2) Figure 5-1 illustrates some of the more common types of false indications, caused by certain handling or cleaning processes.

4. Nonrelevant Indications

Outside the realm of truly false indications there is a category of non-relevant indications, which testing personnel can recognize. These are true indications in the sense that they are caused by surface discontinuities, but the discontinuities are there by design and are in no way a true discontinuity. Most of such nonrelevant indications are easy to recognize since they are related directly to some feature of the assembly that accounts for their presence. Nonrelevant indications include those that appear on articles that are press-fitted, keyed, splined, riveted, or spot welded together and those appearing on castings as a result of loosely adherent scale or a rough surface due to burned-in sand. Such nonrelevant indications, however, must be carefully examined since they may interfere with correct interpretation.

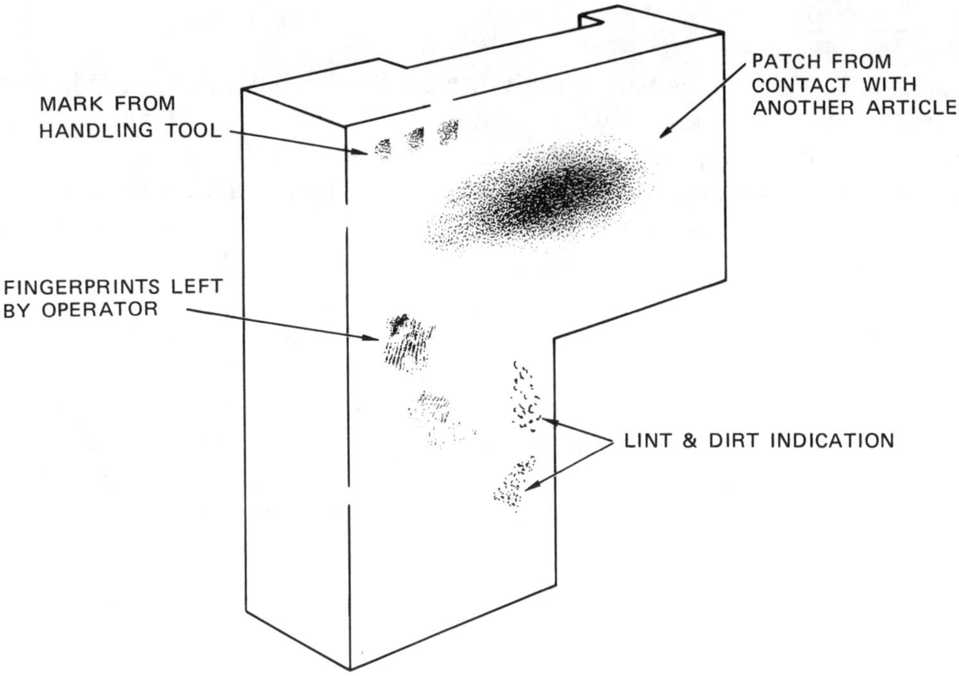

Figure 5-1. Typical False Indications

5. True Indications

True indications are those caused by a discontinuity. The interpretation of an indication as true is a matter of observing the indication, eliminating the possibility of it being a false indication, and then further determining that it is not a nonrelevant indication. Any true indication immediately becomes subject to evaluation as to its cause (type of discontinuity) and the effect of the indicated discontinuity on the service life of the specimen. There are no hard and fast rules that lay down sure methods of determining whether an indication is true. Such definite appraisals require knowledge of the fabrication processes used in creating the article or, in the case of a used article, knowledge of its operational use and the stresses to which it has been subjected.

502 CATEGORIES OF TRUE INDICATIONS

1. General

Discontinuity indications vary widely in appearance but for each indication two interpretive questions must be answered: 1) what type of discontinuity caused the indication, and 2) what is the extent of the discontinuity as evidenced by the extent of the indication. Each indication also requires an answer to the evaluation question: what effect will the indicated discontinuity have on the service life of the article. The answers to the interpretive questions are obtained by observing the indication and identifying the discontinuity from the characteristic appearance of the indication. The answer to the evaluation question is based on a certain knowledge of the seriousness of the discontinuity and complete understanding of the ultimate use of the article. True indications logically fall into five categories: continuous line; intermittent line; rounded; small dots; and diffused or weak. Figure 5-2 illustrates typical true indications.

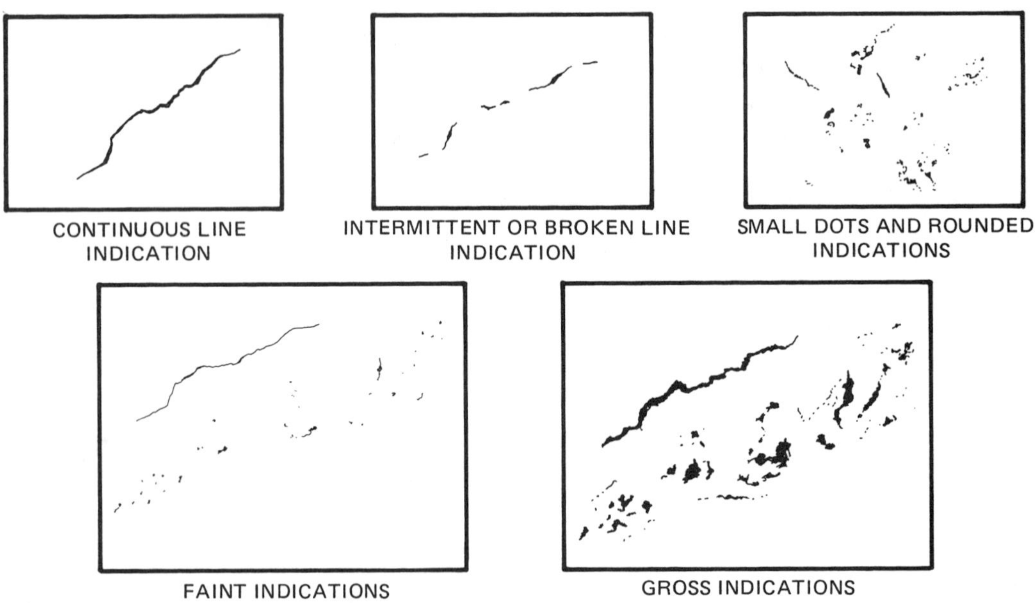

Figure 5-2. Typical True Indications

2. Continuous Line

Continuous line indications are caused by cracks, cold shuts, forging laps, scratches, or die marks. Cracks usually appear as jagged lines; cold shuts as smooth, narrow, straight lines; and forging laps as smooth, wavy lines. Scratches and die marks appear in a variety of linear patterns but are readily recognizable when all penetrant traces are removed, since the bottom of the discontinuity is usually visible.

3. Intermittent Line

The same discontinuities that cause continuous line indications may, under different circumstances, cause intermittent line indications. When an article is worked by grinding, peening, forging, machining, etc., portions of the discontinuities in the surface of the article may be closed by the metal working process. When this occurs, the discontinuities will appear as intermittent lines.

4. Round

Round indications usually are caused by porosity. The porosity may be the result of gas holes, pin holes, or the generally porous makeup of the article. Deep cracks may also appear as round indications since they trap a large amount of penetrant that spreads when the developer is applied. Any round indication that appears singly in an isolated position usually indicates a discontinuity of depth that may or may not be round.

5. Small Dots

Small dot indications result from distontinuities caused by pin holes, by the porous nature of the article, or by excessively coarse grains being used in producing a casting. They may also be the result of cast alloy microshrinkage.

6. Diffused or Weak

Diffused or weak indications are particularly difficult to interpret. Weak indications appearing over a large area are always suspect and when they appear the article is to be thoroughly cleaned and retested. Other weak or diffused indications may be caused by surface porosity but more often are the result of insufficient cleaning, incomplete penetrant removal, or excessive developer.

503 DISCONTINUITY DEPTH DETERMINATION

The vividness of a visible dye indication or the brilliance of a fluorescent indication can sometimes be used as a measure of depth. The greater the depth of a discontinuity the more penetrant it holds and the larger and brighter the indication. Shallow discontinuities entrap only small amounts of penetrant and appear as fine line indications of relatively low brilliance. When evaluation requires more accurate knowledge of the depth of a discontinuity it is often obtained by removing the surface indication and reapplying developer. The subsequent amount and rate of penetrant bleed-out is proportionate to the depth of the discontinuity.

504 TYPICAL INDICATIONS

Figures 5-3 through 5-19 illustrate typical liquid penetrant indications found in liquid penetrant testing.

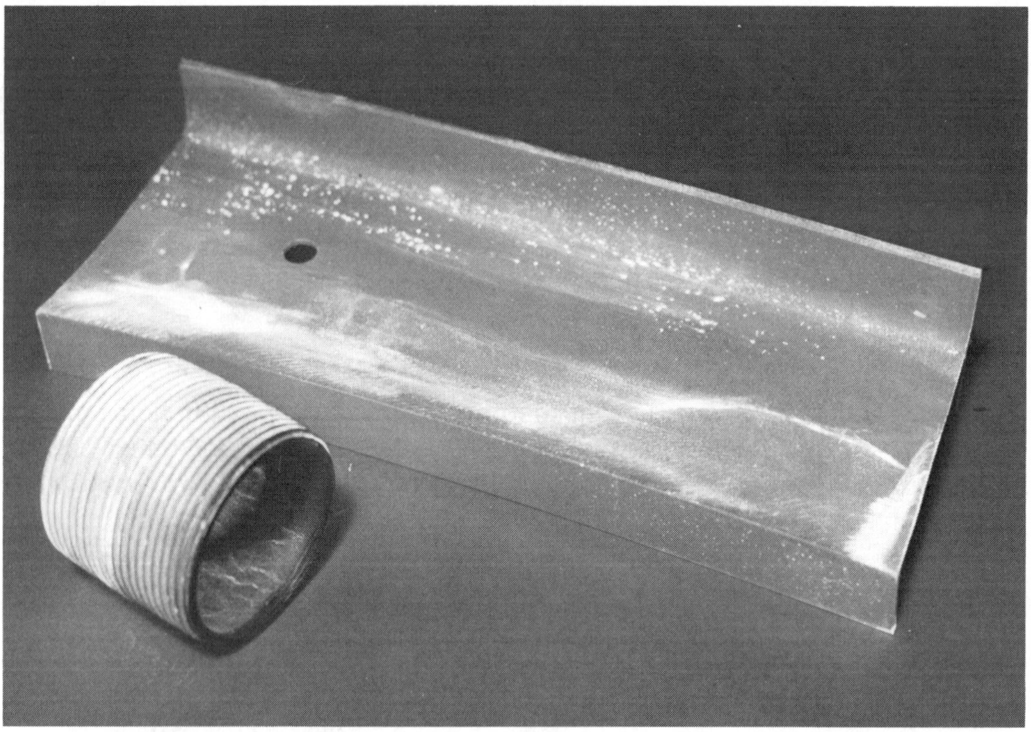

Figure 5-3. Fluorescent Streaks and Patches Due to Incomplete Wash

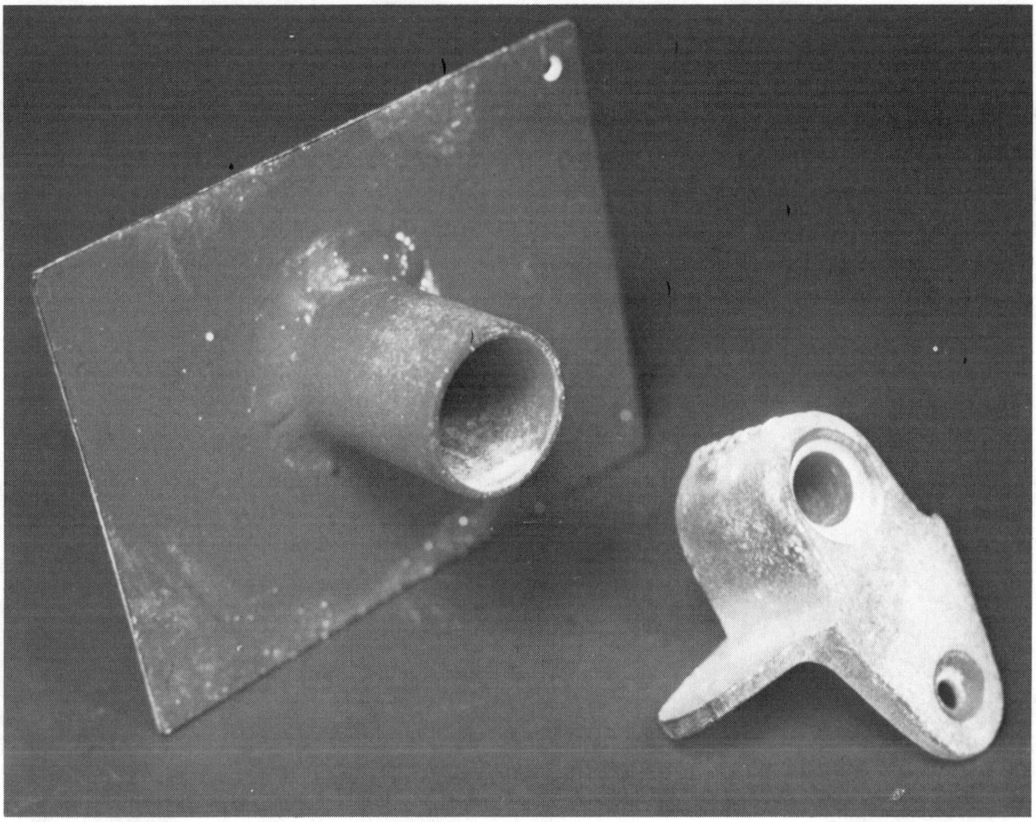

Figure 5-4. Nonrelevant Fluorescent Indications Due to Rough Surfaces and Poor Wash

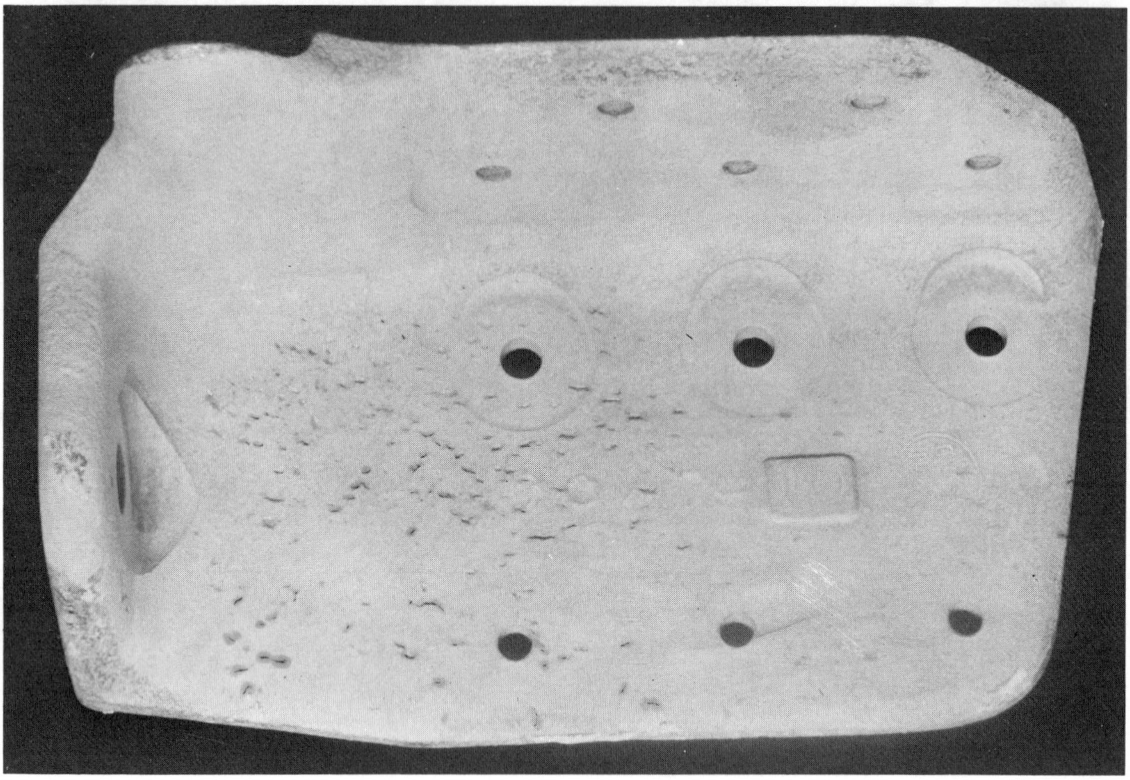

Figure 5-5. Visible Dye Indications of Surface Bursts in Steel Forging

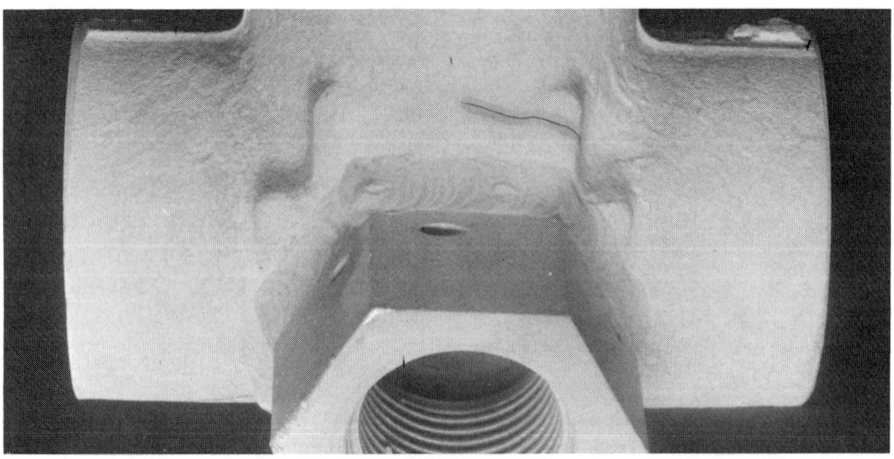

Figure 5-6. Visible Dye Indication of Forging Lap

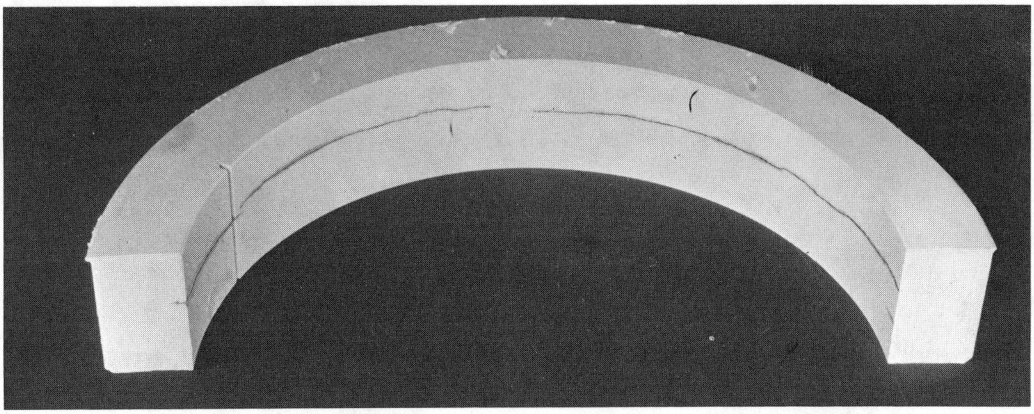

Figure 5-7. Visible Dye Indication of Lamination

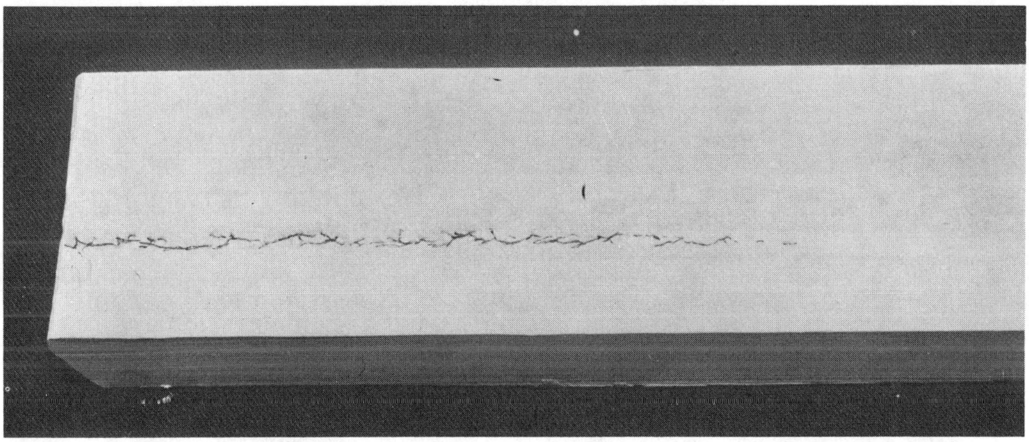

Figure 5-8. Visible Dye Indications of Weldment Shrink Cracks

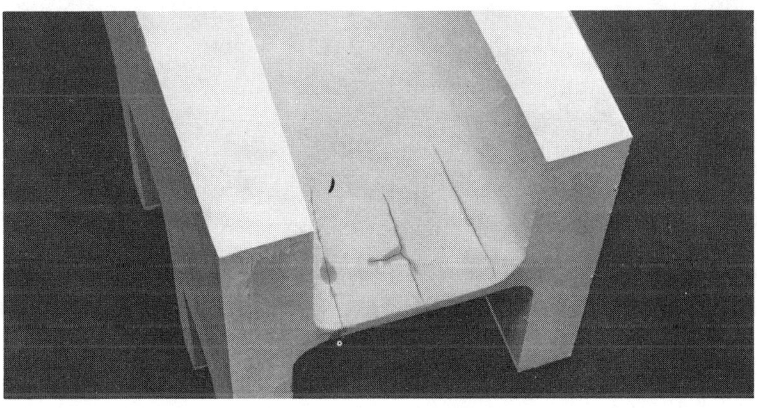

Figure 5-9. Visible Dye Indications of Cracks in Aluminum Forging

Figure 5-10. Visible Dye Indications of Unhealed Porosity in Aluminum Plate

Figure 5-11. Fluorescent Indications of Porosity in Machined Aluminum Plate

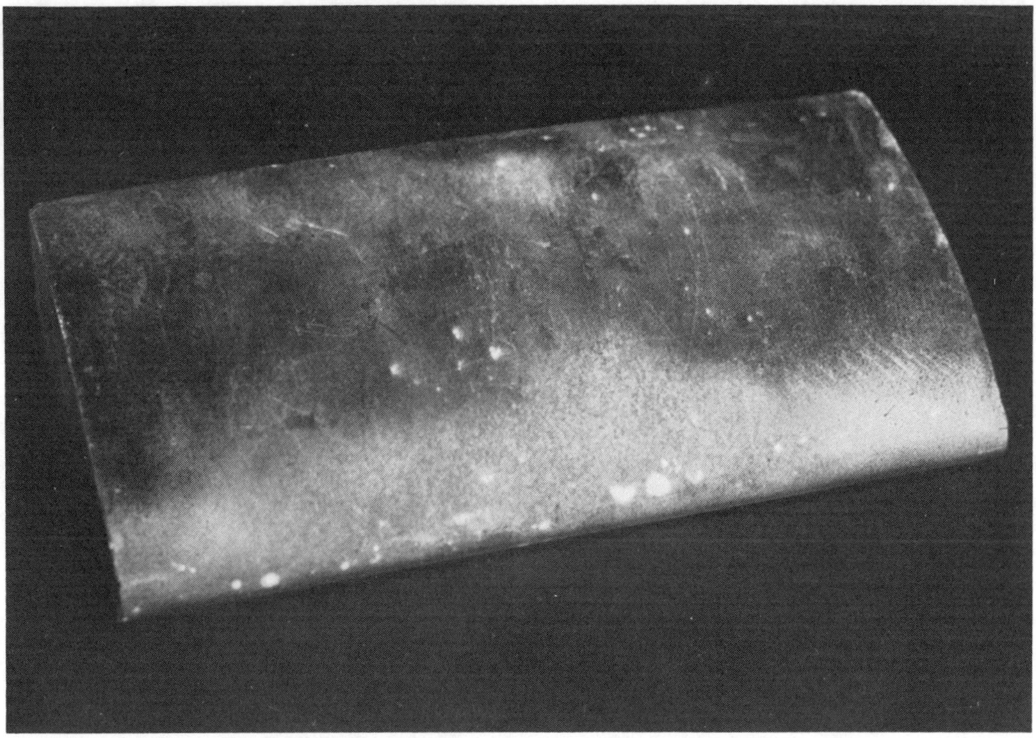

Figure 5-12. Fluorescent Indications of Blow Holes and Random Porosity in Aluminum Casting

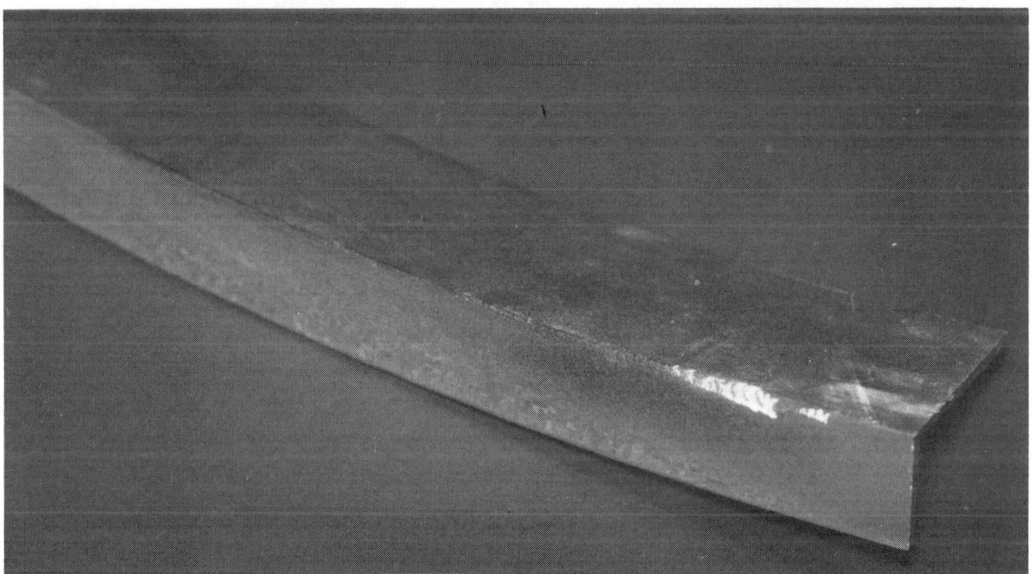

Figure 5-13. Fluorescent Indications of Tears in Stretch Formed Aluminum Angle

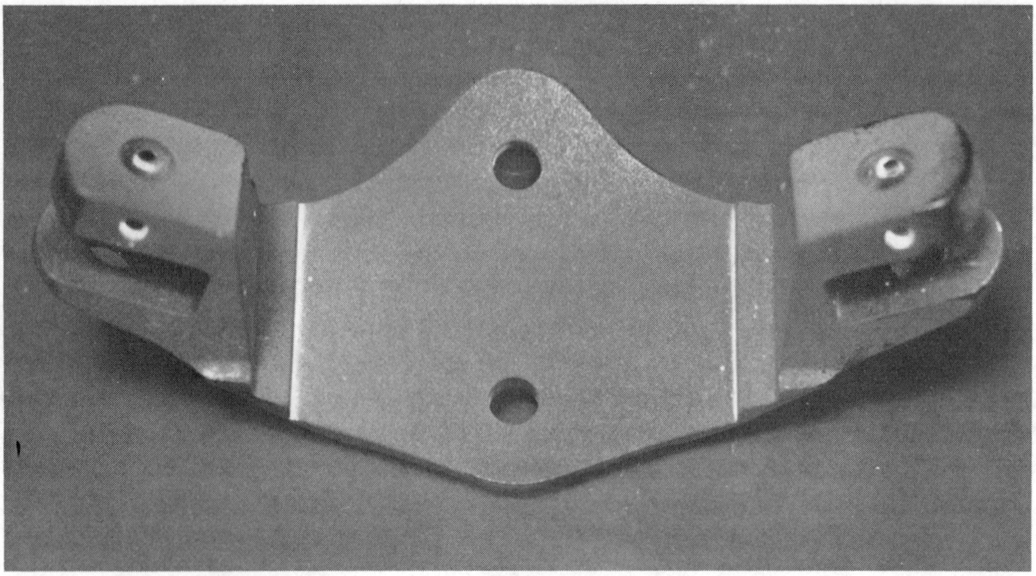

Figure 5-14. Fluorescent Indications of Radius Cracks in Machined Steel Fitting

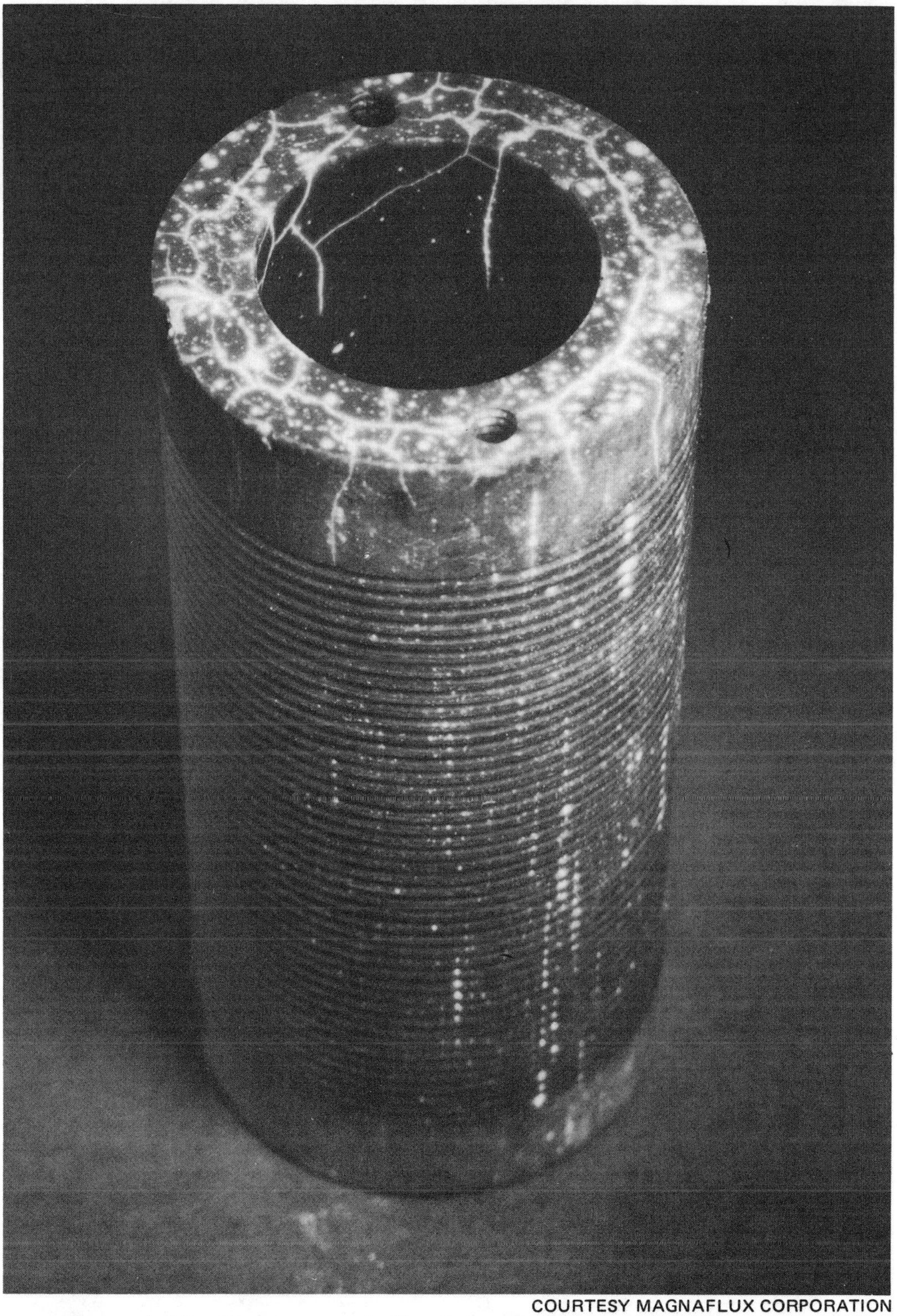

COURTESY MAGNAFLUX CORPORATION

Figure 5-15. Fluorescent Indications of Thermal Cracks, Porosity and Seams in Unglazed Ceramic Coil-Form

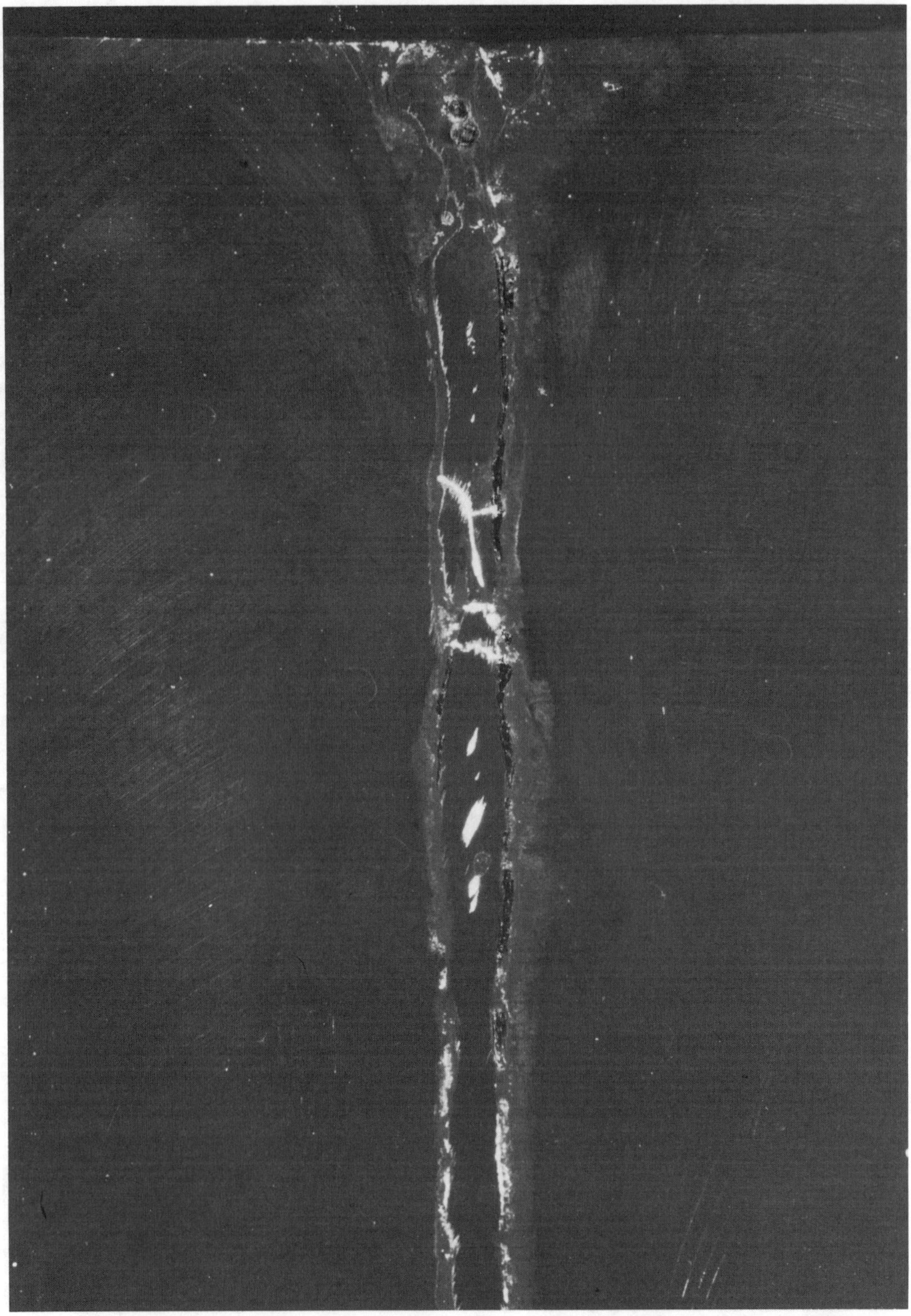

COURTESY MAGNAFLUX CORPORATION

Figure 5-16. Fluorescent Indications of Crater Cracks and Pores in Stainless Steel Weld

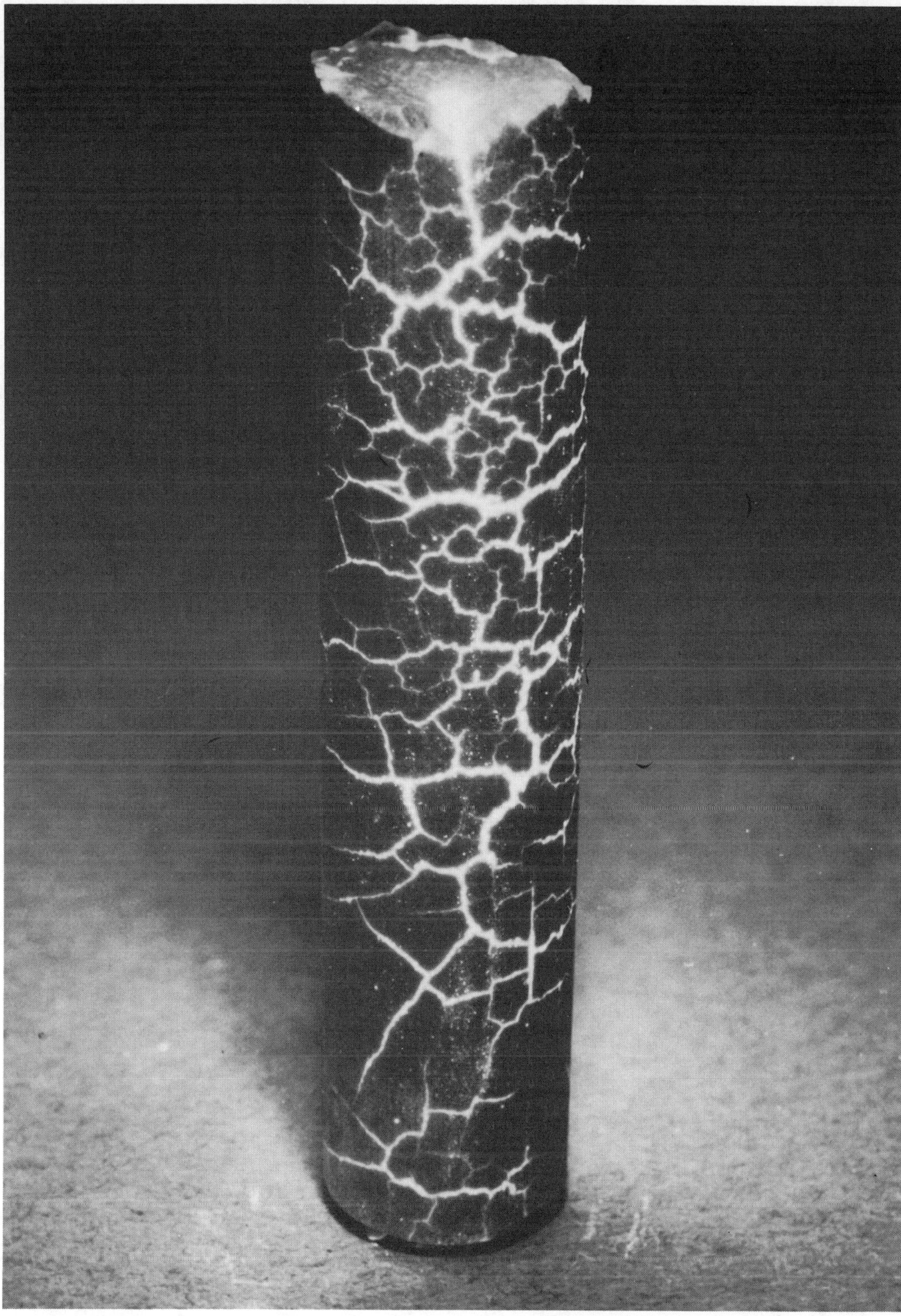

COURTESY MAGNAFLUX CORPORATION

Figure 5-17. Fluorescent Indications of Cracks Produced in Hard-Fired Unglazed Ceramic Rod

COURTESY MAGNAFLUX CORPORATION

Figure 5-18. Fluorescent Indications of Shrinkage Cracks in Unfinished Coupling Castings

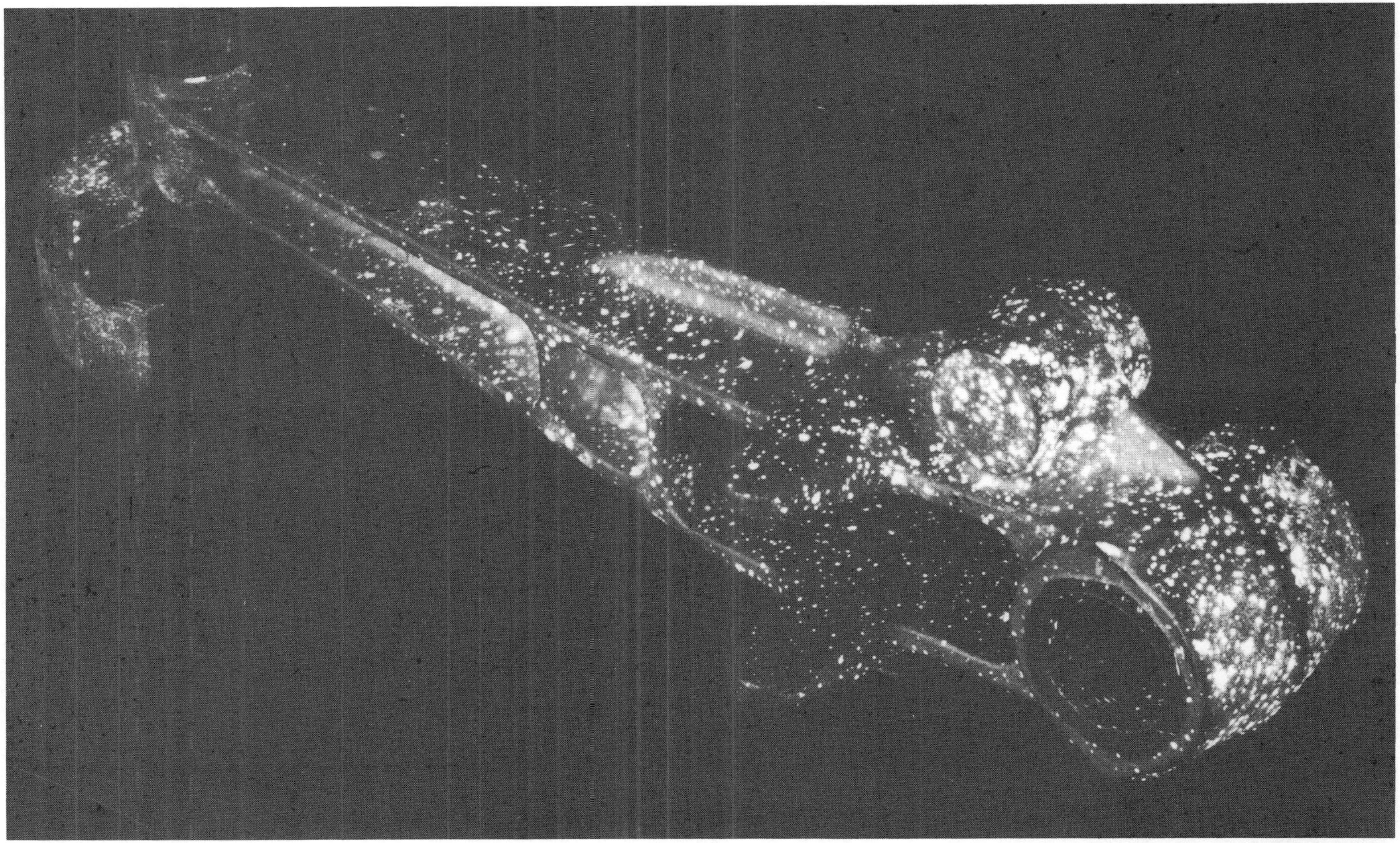

COURTESY MAGNAFLUX CORPORATION

Figure 5-19. Fluorescent Indications of Surface Porosity in Magnesium Sand Casting

CHAPTER 6: QUALITY CONTROL OF PENETRANT TEST MATERIALS

TABLE OF CONTENTS

LIST OF FIGURES

CHAPTER 6: QUALITY CONTROL OF PENETRANT TEST MATERIALS

600 GENERAL

The reliability of any penetrant test is determined in large part by the condition (utility) of the materials used. Even the best procedures are worthless if test materials are faulty. To ensure the satisfactory condition of penetrant test materials, various quality control tests are employed. This chapter discusses primarily the "in-service" checks used to test penetrant materials held in open tanks and subject to contamination or evaporation. Each of the quality control tests mentioned herein is based on the assumption that care, handling, and use of the materials are strictly in accordance with manufacturers' recommendations.

Many additional quality control procedures such as the determination of sulfur and chlorine content, liquid oxygen compatibility, temperature stability, water tolerance, viscosity, flash point, toxicity, tank or shelf life, and developer precipitation rates are described in the various controlling specifications and standards. These procedures however are primarily of interest to the manufacturer or laboratory technician rather than the individual performing or monitoring penetrant tests. Quality control procedures are also readily available from manufacturers and various technical societies such as the American Society for Testing and Materials (ASTM).

601 TEST MATERIAL CONTROL SAMPLES

Since many of the tests used in the control of penetrant materials are comparison tests in which "used" materials are compared with "new" materials, control samples are taken at the time the materials are received from the supplier. These samples are kept in sealed containers and stored where they are not subject to deterioration from heat, light, or evaporation. They are held to be used later in performing comparison tests. The procedures for taking the samples and their renewal are detailed in the controlling specifications and standards.

602 TEST BLOCKS

1. General

The use of test blocks, plates, or panels is often specified in the performance of test procedures used in testing penetrant materials. Depending on the particular test, the materials used in the manufacture of test blocks include aluminum, steel, nickel, glass, and ceramic. Some of the blocks are designed primarily for checking penetrant or "system" sensitivity and performing comparison tests, etc., while others are designed specifically for testing penetrant or emulsifier washability. All, however, are prepared to rigid specifications as detailed in the following paragraphs.

2. Aluminum Test Blocks

Aluminum test blocks measure 2 by 3 inches (50 by 75 mm) and are cut from 5/16-inch thick (8-mm) bare 2024-T3 aluminum alloy plate, with the 3-inch dimension in the direction of rolling. The blocks are heated nonuniformly and water quenched so as to produce thermal cracks. This is accomplished by supporting the block in a frame and heating it with the flame of a gas burner or torch in the center on the underside of the block. The flame remains centered and does not move in any direction during the heating process. A 950 to 980°F (510 to 527°C) Tempilstik, Tempilac, or equivalent, is applied to an area the size of a penny on the top side and directly in the center of the block. The heat of the torch or burner is adjusted so that the block is heated approximately 4 minutes before the Tempilstik or Tempilac melts, after which the block is immediately quenched in cold water (see Figure 6-1). The same operation is then repeated on the other side of the block. A groove approximately 1/16-inch by 1/16-inch deep (1.5 by 1.5 mm) is cut in the 2-inch direction across the center of the heat-affected zone on both sides of the block. This forms two specimen areas on each side of the block and permits the side-by-side application and comparison of two penetrants without cross-contamination. This type of block is widely used for comparing the performance of penetrants under actual crack-finding usage.

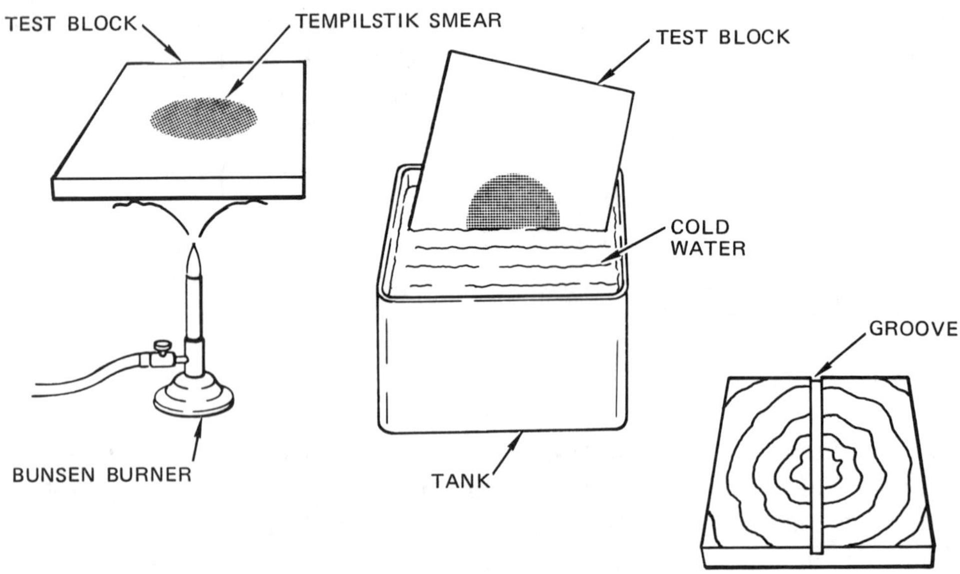

Figure 6-1. Heating and Quenching of Aluminum Test Block

a. Preparation for Use. Prior to use, aluminum test blocks are scrubbed with a bristle brush and liquid solvent, followed by vapor degreasing.

b. Preparation for Reuse. After a test block has been used it is cleaned prior to reuse. The block is heated slowly with a gas burner to 800°F (426°C) as determined by an 800°F Tempilstik, or equivalent, after which the block is quenched in cold water. It is then heated to approximately 225°F (107°C) for 15 minutes to drive off any moisture in the cracks, and is allowed to cool to room temperature.

3. Ceramic Test Blocks

Ceramic test blocks are flat circular disks of unglazed ceramic that, although quite solid and impervious to liquids, have micro-pit surfaces that entrap liquid penetrants. Inasmuch as the micro-pit structure provides a range of pore sizes, a performance comparison can be made of two or more penetrants merely by noting the number or distribution of porosity indications and their brightness in a side-by-side comparison test. Indications appear as a large number of microscopic specks of fluorescence or color, the number increasing as the sensitivity of the penetrant increases.

a. Preparation for Use. The ceramic block is removed from its container (a shallow jar containing alcohol) using a pair of tweezers, or equivalent, to avoid touching with the fingers. The block is then wiped with a soft absorbent tissue, and allowed to dry 2 to 5 minutes.

b. Usage. Using a small applicator, a drop of each penetrant to be tested is applied to the flat surface of the test block. Immediately following application, the penetrants are blotted with a piece of soft tissue by pressing the tissue against the block using a glass platen or other flat object. This prevents the evaporation of the volatile materials in the penetrants from altering indications. The tissue and cover also hold the penetrants in proper contact with the test block and prevent excessive bleeding and possible cross-contamination. Following the required dwell period (usually 10 minutes), the penetrants are processed in accordance with the penetrant manufacturer's recommendations. A developer, however, is not used. In making visual comparisons, both the number of indications observed and the brightness of indications are noted.

c. Cleaning after Use. After use, the ceramic block is placed on edge in alcohol and soaked for several hours or until the penetrant entrapments diffuse into the alcohol. The block is then returned to its storage container, covered with fresh alcohol, and sealed until again needed.

4. Anodized/Plated Test Panels

Stress-cracked anodized aluminum and chrome-plated nickel test panels are frequently employed in determining penetrant sensitivity and washability. The panels are classified according to the size cracks they contain. The grades are; coarse, medium, and fine. The coarse panels provide a low sensitivity level; the medium panels, a medium sensitivity; and the fine panels, a high sensitivity. The panel used depends on the level of sensitivity desired (see Figure 6-2).

COARSE CRACKS   MEDIUM CRACKS   FINE CRACKS

Figure 6-2. Fluorescent Indications of Stress Cracks in Chrome Nickel Test Panels

a. Preparation for Use. Panel surfaces are scrubbed using moderate pressure and a cloth dampened with emulsifier or concentrated soap solution, and followed with a thorough rinsing using a water spray. Following rinsing, the panel is oven dried at 80 to 100°C for 3 to 5 minutes to drive off any moisture remaining in cracks. The panel is then dipped in acetone and agitated for a minute or so, removed, and again oven dried at 60 to 80°C for approximately one minute.

b. Usage. A line is usually drawn along the centerline of the panel and parallel to the lengthwise dimension using a wax pencil or narrow vinyl tape. This forms two specimen areas and permits the side-by-side application and comparison of penetrant materials without cross-contamination.

5. Aluminum Test Panels

Aluminum test plates or panels are used in testing the washability of water-washable visible dye penetrants. The panels may be of any convenient size, 2 by 4 inches (50 by 100 mm) or larger, and are cut from clad aluminum alloy. The test surfaces are used in the "as-rolled" condition. Prior to use, the panel is degreased, after which it is heated to approximately 225°F (107°C), then cooled to room temperature in a dessicator.

6. Steel Test Panels

Steel test plates or panels are used in testing the washability of water-washable fluorescent penetrants, and visible dye or fluorescent penetrants involving the use of an emulsifier or solvent remover. The panels are prepared from annealed type 301 or 302 stainless steel and measure 2 by 4 inches (50 by 100 mm) or larger. Each is sandblasted on one side with 100-mesh average size grit, using 60 pounds air pressure (414 kPa), with the gun held approximately 18 inches (457 mm) from the surface. The sandblasting is continued until a uniform matte surface is obtained. It is the sandblasted surface that is used in performing tests.

a. Preparation for Use. Prior to use, the panels are cleaned by vapor degreasing, heated to 225°F (107°C), and then allowed to cool to room temperature in a desiccator.

b. Cleaning for Reuse. After each use, the panels are vapor degreased and resandblasted.

7. Glass Plates

One-quarter-inch thick (6 mm) glass plates (black for fluorescent penetrants and clear for visible dye penetrants) are also used to test the washability of penetrants. They are fabricated by sandblasting sheets of glass using 100-mesh average size grit at 60 psi air pressure (414 kPa), so as to provide a uniform matte surface.

603 PENETRANT TESTS

1. General

The "in-service" quality of the penetrants used in liquid penetrant testing is determined by a check of penetrant sensitivity, water content, contamination, and washability. The tendency toward fading of the penetrant dyes is also checked by a simple comparison test. The tests described in the following paragraphs are typical of those performed on "used" or questionable penetrants. As a general rule the tests referenced in paragraphs 603.4 and 603.5 would be performed by laboratory technicians in a laboratory and not by the liquid penetrant testing specialist.

2. Sensitivity Test

When performing the sensitivity test, the penetrant to be tested is applied to one half of the test block (Para. 602.2, 602.3, or 602.4) and the new or reference penetrant (Para. 601) is applied to the remaining half. The processing used – including dwell time, emulsification or removal, and developing – is that recommended by the penetrant manufacturer. The indications are then visually compared under the appropriate lighting (normal or white light for visible dye indications and black light for fluorescent indications). If a noticeable difference exists in the sensitivity or brightness of indications (as determined by visual observation), the penetrant is discarded. Likewise, if the penetrant is off-color or shows evidence of contamination from dirt or lint, it is discarded.

Coincidently with the visual examination of fluorescent indications, the black light source used to view indications must have an intensity of at least 800 microwatts per square centimeter at the test surface.

3. Water Washability Test

To achieve maximum contrast between indications and background, excess surface penetrant must be readily removable. In the water washability test, the washability performance of the penetrant being tested is compared to that of the reference penetrant (Para. 601). The penetrants are applied to separate test panels (Para. 602.5, 602.6, or 602.7). After normal dwell and draining periods, and emulsification (if applicable), the penetrants are washed from the panels using a uniform water spray. If washing is found to be difficult, or retention of background dye is noticeably different from that of the reference penetrant, the penetrant is discarded.

4. Water Content Test

Water content of a penetrant is best determined by the test described in ASTM Standard D-95. One hundred milliliters (ml) of the penetrant is placed in a boiling flask with a similar quantity of moisture-free xylene. The flask is connected to a reflux condenser so that the condensate drops into a 25-ml graduated tube where the water settles out. When no more water is being gathered in the graduated tube (usually after a period of an hour) the boiling process is terminated. After cooling, the volume of water in the graduated tube is read. The volume in milliliters is the percent of water by volume present in the penetrant.

The penetrant is also examined for evidence of gelling, separating, clouding, coagulating, or floating of water on the penetrant surface. If any of these conditions exist, or the percent of water exceeds specification requirements, the penetrant is discarded.

5. Fluorescent Brightness Test

The fluorescent brightness test is usually performed with a modified photofluorometer. In this test, a small amount of the penetrant to be tested and the reference penetrant (Para. 601) are diluted with a nonfluorescent highly volatile solvent such as methylene chloride. Test papers, cut to fit the sample holder of the photofluorometer, are then dipped into the solutions, withdrawn, and allowed to air dry. Following drying, the samples are placed in a preheated oven at 225°F (107°C) for 5 minutes. Five samples are prepared of the penetrant being tested and six of the reference penetrant (the extra reference penetrant sample is used as the master for setting up the photofluorometer). The samples are then alternately read on the photofluorometer and the results compared. If the fluorescent brightness of the penetrant being tested should drop below 85% of the reference penetrant, the penetrant is discarded.

604 EMULSIFIER TESTS

1. General

Emulsifiers are usually tested for their sensitivity, washability, water content, and the amount of contamination from penetrants. As a general rule the tests referenced in paragraphs 604.4 and 604.5 would be performed by laboratory technicians and not by the liquid penetrant testing specialist.

2. Sensitivity Test

The emulsifier sensitivity test is identical to that used for penetrants and described in paragraph 603.2.

3. Water Washability Test

The water washability test is identical to the test used for penetrants and described in paragraph 603.3.

4. Water Content Test

The water content test is identical to the test used for penetrants and described in paragraph 603.4.

5. Contamination Test

The principal contaminant of emulsifiers held in open tanks is penetrant. To limit this contamination, a fluorescence test is periodically performed. In this test, two blends of emulsifier and penetrant are prepared. The first blend consists of 75% new emulsifier and 25% new penetrant (Para. 601). The second blend consists of the same percentages of used emulsifier and penetrant. The samples are applied to separate test panels (Para. 602.6) and permitted to drain at a 45-degree angle for 15 minutes.

The "used" penetrant-emulsifier test panel is then dipped momentarily in the emulsifier being tested; the reference panel is dipped in the reference emulsifier. The dwell period used will depend on the emulsifier but it usually ranges from 3 to 30 seconds. Then, using a special spray apparatus and a 20-grain hard water solution and 5 pounds air pressure (34 kPa), the panels are rinsed for 30 seconds with the spray nozzle held 90 degrees to the surface and 12 inches (305 mm) away.

Following the rinse, the panels are dried in a warm air dryer, after which (if applicable) the appropriate developer is applied. The panels are then compared under black light. Should the residual fluorescence of the used penetrant-emulsifier combination noticeably exceed that of the reference sample (by visual observation), the emulsifier is discarded.

605 DEVELOPER TESTS

1. Dry Developer

Dry developers used in open tanks are usually tested only by observation. Since they are not hydroscopic, they do not absorb moisture from the air, and are relatively trouble-free if they do not come in contact with water. Any dry developer that is found lumpy or caked instead of light and fluffy, or that shows any other sign of having been wet, is discarded. Additionally the developer is visually examined for dirt. It is also checked under black light for fluorescent dye contamination. Should either condition exist, the developer is discarded.

2. Wet Developer

Wet (aqueous) developers are usually tested only for proper density and possible contamination from dirt or penetrant. Specific gravity is measured with a hydrometer (see Figure 3-6). If reading differs from specification requirements (usually 1.020 to 1.028), either powder or vehicle is added to the developer in sufficient quantities to bring the density within acceptable limits. As a general rule this test would be performed by laboratory technicians and not by the liquid penetrant testing specialist. Additionally, a small sample of the developer is taken from the tank and visually examined for dirt. It is also checked for fluorescent dye contamination under black light. If either condition is in evidence, the developer is discarded.

CHAPTER 7: COMPARISON AND SELECTION OF NDT PROCESSES

TABLE OF CONTENTS

LIST OF FIGURES

CHAPTER 7: COMPARISON AND SELECTION OF NDT PROCESSES

700 GENERAL

This chapter summarizes the characteristics of various types of discontinuities, and lists the NDT methods that may be employed to detect each type of discontinuity.

The relationship between the various NDT methods and their capabilities and limitations when applied to the detection of a specific discontinuity is shown. Such variables as type of discontinuity (inherent, process, or service), manufacturing processes (heat treating, machining, welding, grinding, or plating), and limitations (metallurgical, structural, or processing) also help in determining the sequence of testing and the ultimate selection of one test method over another.

701 METHOD IDENTIFICATION

Figures 7-1 through 7-5 illustrate five NDT methods. Each illustration shows the three elements involved in all five tests, the different methods in each test category, and tasks that may be accomplished with a specific method.

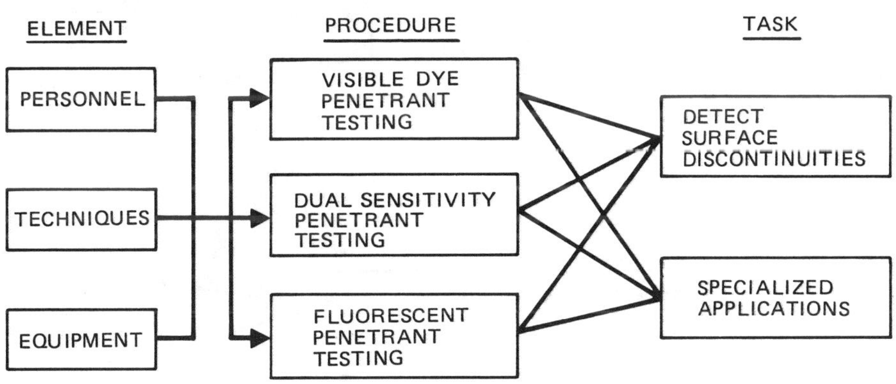

Figure 7-1. Liquid Penetrant Test

702 NDT DISCONTINUITY SELECTION

The discontinuities that are discussed in paragraphs 706 through 732 are only some of the many hundreds that are associated with the various products of today's industry. During the selection of discontinuities for inclusion in this chapter, only those discontinuities which would not be radically changed under different conditions of design, configuration, standards, and environment were chosen.

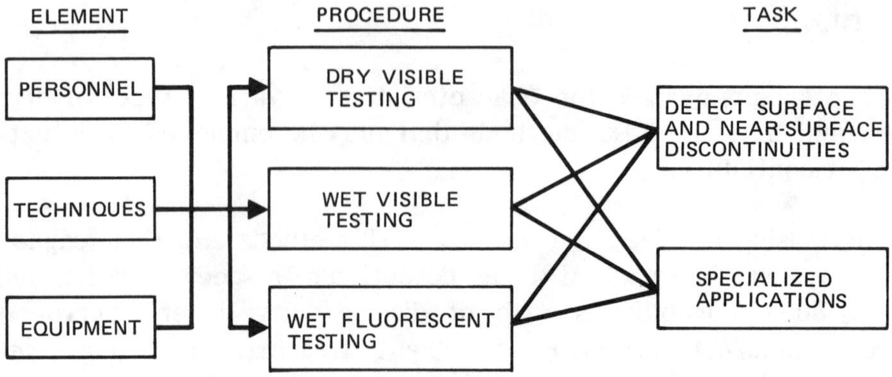

Figure 7-2. Magnetic Particle Test

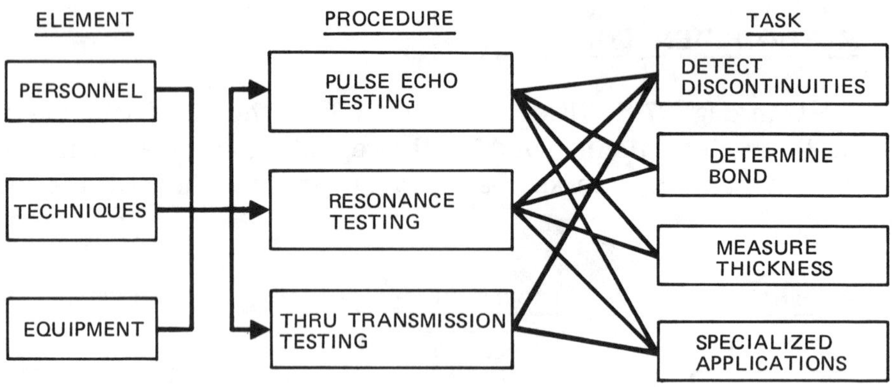

Figure 7-3. Ultrasonic Test

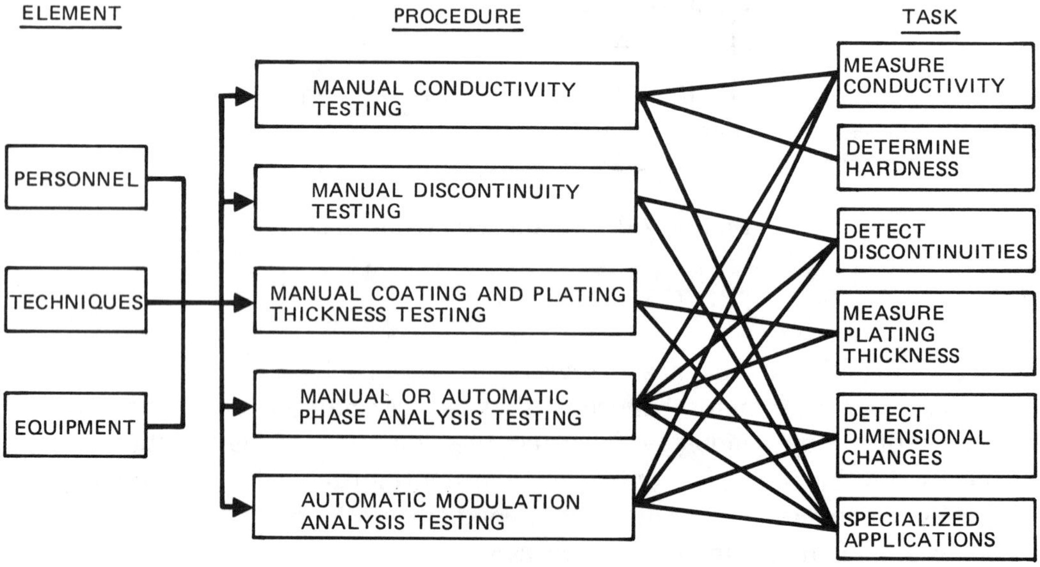

Figure 7-4. Eddy Current Test

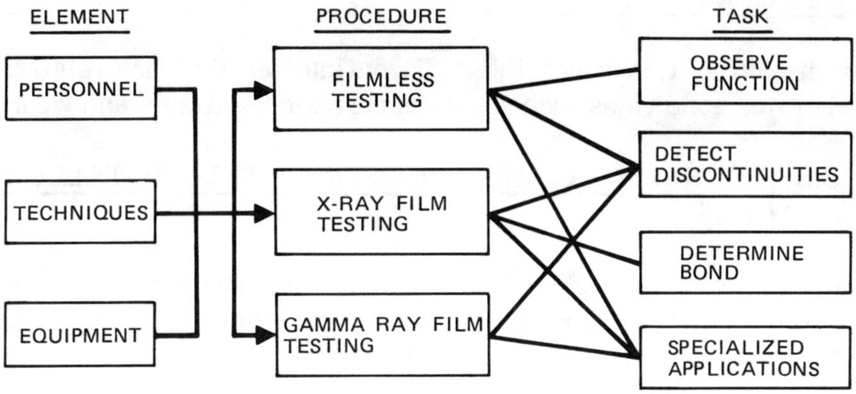

Figure 7-5. Radiographic Test

703 DISCONTINUITY CATEGORIES

Each of the specific discontinuities are divided into three general categories: inherent, processing, and service. Each of these categories is further classified as to whether the discontinuity is associated with ferrous or nonferrous materials, the specific material configuration, and the manufacturing processes if applicable.

1. Inherent Discontinuities

Inherent discontinuities are those discontinuities that are related to the solidification of the molten metal. There are two types.

a. Wrought. Inherent wrought discontinuities cover those discontinuities which are related to the melting and original solidification of the metal or ingot.

b. Cast. Inherent cast discontinuities are those discontinuities which are related to the melting, casting, and solidification of the cast article. It includes those discontinuities that would be inherent to manufacturing variables such as inadequate feeding, gating, excessively high pouring temperature, entrapped gases, handling, and stacking.

2. Processing Discontinuities

Processing discontinuities are those discontinuities that are related to the various manufacturing processes such as machining, forming, extruding, rolling, welding, heat treating, and plating.

3. Service Discontinuities

Service discontinuities cover those discontinuities that are related to the various service conditions such as stress corrosion, fatigue, and wear.

704 DISCONTINUITY CHARACTERISTICS AND METALLURGICAL ANALYSIS

"Discontinuity characteristics," as used in this chapter, encompasses an analysis of specific discontinuities and references actual photos that illustrate examples of the discontinuity. The discussions cover:

a. Origin and location of discontinuity (surface, near surface, or internal).

b. Orientation (parallel or normal to the grain).

c. Shape (flat, irregularly shaped, or spiral).

d. Photo (micrograph and/or typical overall view of the discontinuity).

e. Metallurgical analysis (how the discontinuity is produced and at what stage of manufacture).

705 NDT METHODS APPLICATION AND LIMITATIONS

1. General

The technological accomplishments in the field of nondestructive testing have brought test reliability and reproducibility to a point where the design engineer may now selectively zone the specific article. Zoning is based upon the structural application of the end product and takes into consideration the environment as well as the loading characteristics of the article. Such an evaluation in no way reduces the end reliability of the product, but evaluation does reduce needless rejection of material that otherwise would have been acceptable.

Just as the structural application within the article varies, the allowable discontinuity size will vary depending on the configuration and method of manufacture. For example, a die forging that has large masses of material and extremely thin web sections would not require the same level of acceptance over the entire forging. The forging can be zoned for rigid control of areas where the structural loads are higher, and less rigid for areas where the structural loads permit larger discontinuities.

The nondestructive testing specialist must also select the method which will satisfy the design objective of the specific article and not assume that all NDT methods can produce the same reliability for the same type of discontinuity.

2. Selection of the NDT Method

In selecting the NDT method for the evaluation of a specific discontinuity keep in mind that NDT methods may supplement each other and that several NDT methods may be capable of performing the same task. The selection of one method over another is based upon such variables as:

a. Type and origin of discontinuity

b. Material manufacturing processes

c. Accessibility of article

d. Level of acceptability desired

e. Equipment available

f. Cost.

A planned analysis of the task must be made for each article requiring NDT testing.

The NDT methods listed for each discontinuity in paragraphs 706 through 732 are in order of preference for that particular discontinuity. However, when reviewing the discussions, it should be kept in mind that rapidly developing new techniques in the NDT field may alter the order of test preference.

3. Limitations

The limitations applicable to the various NDT methods will vary with the applicable standard, the material, and the service environment. Limitations not only affect the NDT method but, in many cases, also affect the structural reliability of the test article. For these reasons, limitations that are listed for one discontinuity may also be applicable to other discontinuities under slightly different conditions of material or environment. In addition, the many combinations of environment, location, material, and test capability do not permit mentioning all limitations that may be associated with the problems of locating a specific discontinuity. The intent of this chapter is fulfilled if you are made aware of the many factors that influence the selection of a valid NDT method.

706 BURST

1. Category. Processing

2. Material. Ferrous and Nonferrous Wrought Material

3. Discontinuity Characteristics

Surface or internal. Straight or irregular cavities varying in size from wide open to very tight. Usually parallel with the grain. Found in wrought material that required forging, rolling, or extruding. (See Figure 7-6.)

4. Metallurgical Analysis

a. Forging bursts are surface or internal ruptures caused by processing at too low a temperature, excessive working, or metal movement during forging, rolling, or extruding operation.

b. A burst does not have a spongy appearance and, therefore, is distinguishable from a pipe, even when it occurs at the center.

c. Bursts are often large and are very seldom healed during subsequent working.

5. NDT Methods Application and Limitations

a. Ultrasonic Testing Method.

(1) Normally used for the detection of internal bursts.

(2) Bursts are definite breaks in the material and resemble a crack, producing a very sharp reflection on the scope.

(3) Ultrasonic testing is capable of detecting varying degrees of burst, a condition not detectable by other NDT methods.

(4) Nicks, gouges, raised areas, tool tears, foreign material, or gas bubbles on the article may produce adverse ultrasonic test results.

b. Eddy Current Testing Method. Not normally used. Testing is restricted to wire, rod, and other articles under 0.250 inch (6.35 mm) diameter.

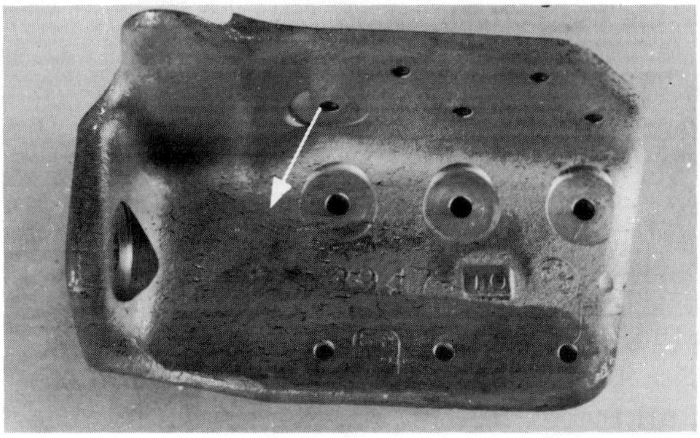

A. FORGING EXTERNAL BURST

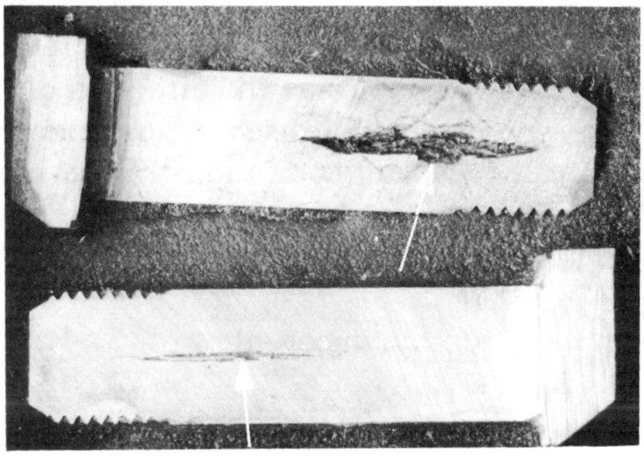

B. BOLT INTERNAL BURST

C. ROLLED BAR INTERNAL BURST

D. FORGED BAR INTERNAL BURST

Figure 7-6. Burst Discontinuities

c. Magnetic Particle Testing Method.

(1) Usually used on wrought ferromagnetic material in which the burst is open to the surface or has been exposed to the surface.

(2) Results are limited to surface and near surface evaluation.

d. Liquid Penetrant Testing Method. Not normally used. When fluorescent penetrant is to be applied to an article previously dye penetrant tested, all traces of dye penetrant should first be removed by prolonged cleaning in applicable solvent.

e. Radiographic Testing Method. Not normally used. Such variables as the direction of the burst, close interfaces, wrought material, discontinuity size, and material thickness restrict the capability of radiography.

707 COLD SHUTS

1. Category. Inherent

2. Material. Ferrous and Nonferrous Cast Material

3. Discontinuity Characteristics

Surface and subsurface. Generally appear as smooth indentations on the cast surface resembling a forging lap. (See Figure 7-7.)

4. Metallurgical Analysis

Cold shuts are produced during casting of molten metal. They may result from splashing, surging, interrupted pouring, or the meeting of two streams of metal coming from different directions. Cold shuts are also caused by the solidification of one surface before other metal flows over it, the presence of interposing surface films on cold, sluggish metal, or any factor that prevents fusion where two surfaces meet. Cold shuts are more prevalent in castings formed in a mold having several sprues or gates.

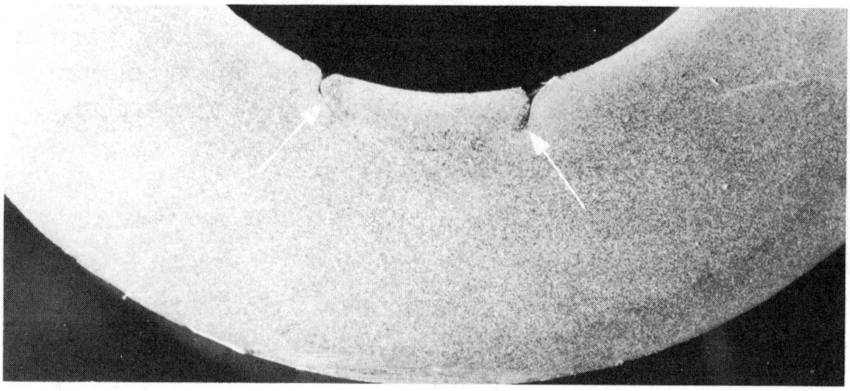

A. SURFACE COLD SHUT

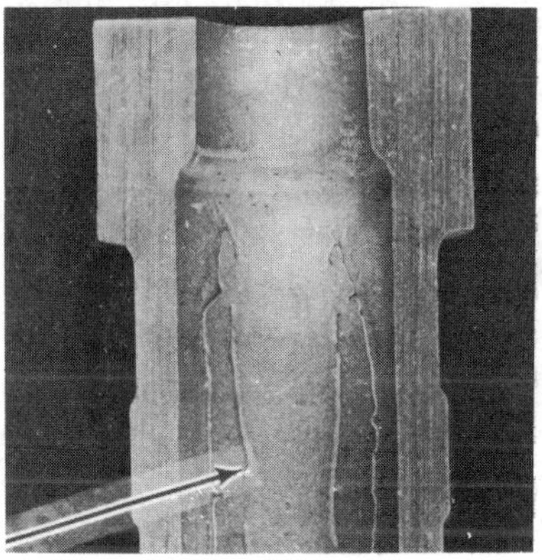

B. INTERNAL COLD SHUT

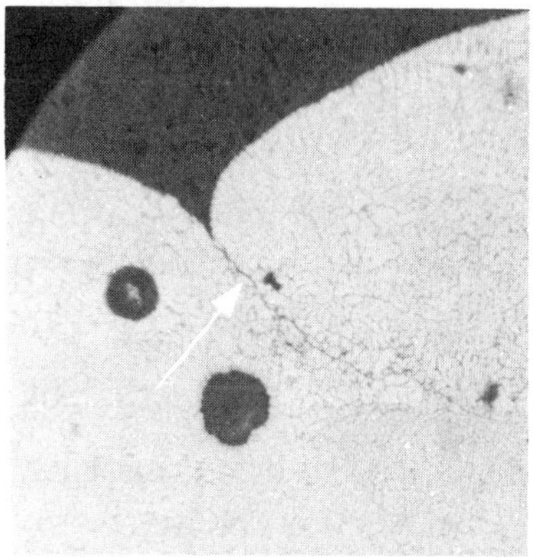

C. SURFACE COLD SHUT MICROGRAPH

Figure 7-7. Cold Shut Discontinuities

5. NDT Methods Application and Limitations

a. Liquid Penetrant Testing Method.

(1) Normally used to evaluate surface cold shuts in both ferrous and nonferrous materials.

(2) Indications appear as a smooth, regular, continuous or intermittent line.

(3) Liquid penetrants used to test nickel base alloys, certain stainless steels, and titanium should not exceed 1% sulfur or chlorine.

(4) Certain castings may have surfaces that are blind and from which removal of excess penetrant may be difficult.

(5) The geometric configuration (recesses, orifices, and flanges) of a casting may permit buildup of wet developer thereby masking any detection of a discontinuity.

b. Magnetic Particle Testing Method.

(1) Normally used for the evaluation of ferromagnetic materials.

(2) The metallurgical nature of 431 corrosion-resistant steel is such that, in some cases, magnetic particle testing indications are obtained which do not result from a crack or other harmful discontinuities. These indications arise from a duplex structure within the material, wherein one portion exhibits strong magnetic retentivity and the other does not.

c. Radiographic Testing Method.

(1) Cold shuts are normally detectable by radiography while testing for other casting discontinuities.

(2) Cold shuts appear as a distinct dark line, or band, of variable length and width, and definite smooth outline.

(3) The casting configuration may have inaccessible areas that can only be tested by radiography.

d. Ultrasonic Testing Method. Not recommended. Cast structure and article configuration do not, as a general rule, lend themselves to ultrasonic testing.

e. Eddy Current Testing Method. Not recommended. Article configuration and inherent material variables restrict the use of this method.

708 FILLET CRACKS (BOLTS)

1. Category. Service

2. Material. Ferrous and Nonferrous Wrought Material

3. Discontinuity Characteristics

Surface. Located at the junction of the fillet with the shank of the bolt and progressing inward. (See Figure 7-8.)

A. FILLET FATIGUE FAILURE

B. FRACTURE AREA OF (A) SHOWING TANGENCY POINT OF FAILURE

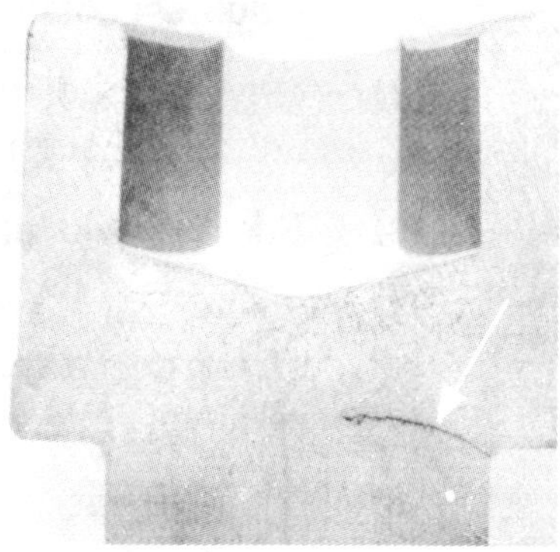

C. CROSS-SECTIONAL AREA OF FATIGUE CRACK IN FILLET SHOWING TANGENCY POINT IN RADIUS

Figure 7-8. Fillet Crack Discontinuity

4. Metallurgical Analysis

Fillet cracks occur where a marked change in diameter occurs, such as at the head-to-shank junction where stress risers are created. During the service life of a bolt repeated loading takes place whereby the tensile load fluctuates in magnitude due to the operation of the mechanism. These tensile loads can cause fatigue failure starting at the point where the stress risers occur. Fatigue failure, which is surface phenomenon, starts at the surface and propagates inward.

5. NDT Methods Application and Limitations

a. Ultrasonic Testing Method.

(1) Used extensively for service associated discontinuities of this type.

(2) A wide selection of transducers and equipment enable on-the-spot evaluation for fillet crack.

(3) Since fillet cracks are a definite break in the material, the scope pattern will be a very sharp reflection. (Propagation can be monitored by using ultrasonics.)

(4) Ultrasonic equipment has extreme sensitivity, and established standards should be used to give reproducible and reliable results.

b. Liquid Penetrant Testing Method.

(1) Normally used during in-service overhaul or troubleshooting.

(2) May be used for both ferromagnetic and nonferromagnetic bolts, although usually confined to the nonferromagnetic.

(3) Fillet cracks appear as sharp, clear indications.

(4) Structural damage may result from exposure of high-strength steels to paint strippers, alkaline coating removers, deoxidizer solutions, etc.

(5) Entrapment of penetrant under fasteners, in holes, under splices, and in similar areas may cause corrosion due to the penetrant's affinity for moisture.

c. Magnetic Particle Testing Method.

(1) Only used on ferromagnetic bolts.

(2) Fillet cracks appear as sharp clear indications with a heavy buildup.

(3) Sharp fillet areas may produce nonrelevant magnetic indications.

(4) 17.7 pH steel is only slightly magnetic in the annealed condition, but becomes strongly magnetic after heat treatment, when it may be magnetic particle tested.

d. Eddy Current Testing Method. Not normally used for detection of fillet cracks. Other NDT methods are more compatible to the detection of this type of discontinuity.

e. Radiographic Testing Method. Not normally used for detection of fillet cracks. Surface discontinuities of this type would be difficult to evaluate due to size of crack in relation to the thickness of material.

709 GRINDING CRACKS

1. Category. Processing

2. Material. Ferrous and Nonferrous

3. Discontinuity Characteristics

Surface. Very shallow and sharp at the root. Similar to heat-treat cracks and usually, but not always, occur in groups. Grinding cracks generally occur at right angles to the direction of grinding. They are found in highly heat-treated articles, chrome plated, case hardened, and ceramic materials that are subjected to grinding operations. (See Figure 7-9.)

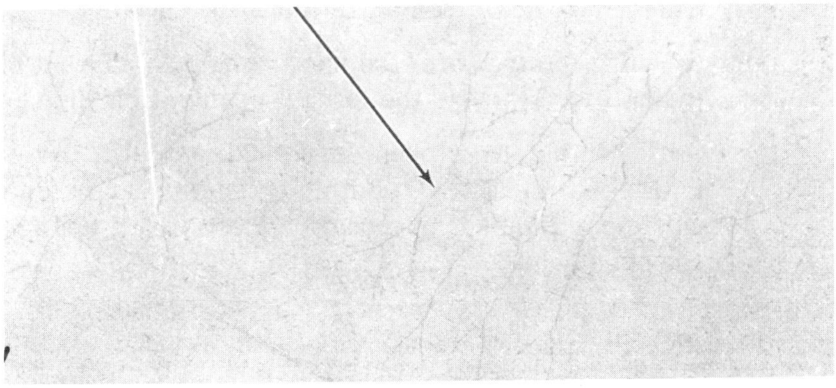

A. TYPICAL CHECKED GRINDING CRACK PATTERN

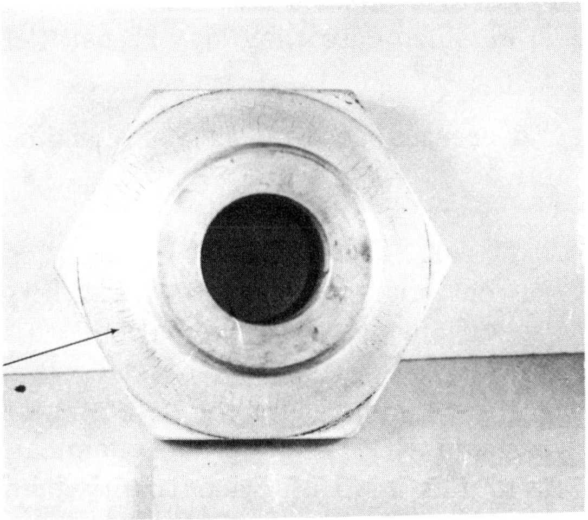

B. GRINDING CRACK PATTERN NORMAL TO GRINDING

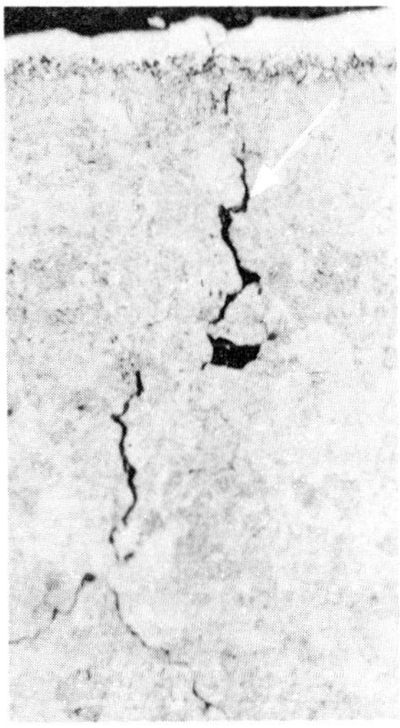

C. MICROGRAPH OF GRINDING CRACK

Figure 7-9. Grinding Crack Discontinuity

4. Metallurgical Analysis

Grinding of hardened surfaces frequently introduces cracks. These thermal cracks are caused by local overheating of the surface being ground. The overheating is usually caused by lack of, or poor, coolant; a dull, or improperly ground, wheel; too rapid feed; or too heavy cut.

5. NDT Methods Application and Limitations

a. Liquid Penetrant Testing Method.

(1) Normally used on both ferrous and nonferrous materials for the detection of grinding cracks.

(2) Liquid penetrant indication will appear as irregular, checked, or scattered pattern of fine lines.

(3) Grinding cracks are the most difficult discontinuity to indicate and require the longest penetration time.

(4) Articles that have been degreased may still have solvent entrapped in the discontinuity and should be allowed sufficient time for evaporation prior to the application of the penetrant.

b. Magnetic Particle Testing Method.

(1) Restricted to ferromagnetic materials.

(2) Grinding cracks generally occur at right angles to grinding direction, although in extreme cases a complete network of cracks may appear, in which case they may be parallel to the magnetic field.

(3) Magnetic sensitivity decreases as the size of grinding crack decreases.

c. Eddy Current Testing Method. Not normally used for detection of grinding cracks. Eddy current equipment has the capability and can be developed for a specific nonferrous application.

d. Ultrasonic Testing Method. Not normally used for detection of grinding cracks. Other forms of NDT are more economical, faster, and better adapted to this type of discontinuity than ultrasonics.

e. Radiographic Testing Method. Not recommended for detection of grinding cracks. Grinding cracks are too tight and small. Other NDT methods are more suitable for detection of grinding cracks.

710 CONVOLUTION CRACKS

1. Category. Processing

2. Material. Nonferrous

3. Discontinuity Characteristics

Surface. Range in size from microfractures to open fissures. Situated on the periphery of the convolutions and extend longitudinally in direction of rolling. (See Figure 7-10.)

4. Metallurgical Analysis

The rough "orange peel" effect of convolution cracks is the result of either a forming operation that stretches the material or from chemical attack such as pickling treatment. The roughened surface contains small pits that form stress risers. Subsequent service application (vibration and flexing) may introduce stresses that act on these pits and form fatigue cracks as shown in Figure 7-10.

5. NDT Methods Application and Limitations

a. Radiographic Testing Method.

(1) Used extensively for this type of failure.

(2) The configuration of the article and the location of the discontinuity limits detection almost exclusively to radiography.

(3) Orientation of convolutions to X-ray source is very critical since those discontinuities that are not normal to X-ray may not register on the film due to the small change in density.

(4) Liquid penetrant and magnetic particle testing may supplement but not replace radiographic and ultrasonic testing.

A. TYPICAL CONVOLUTION DUCTING

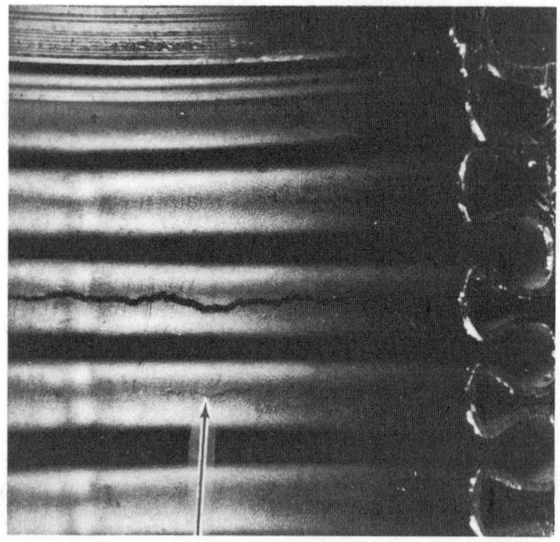

B. CROSS-SECTION OF CRACKED CONVOLUTION

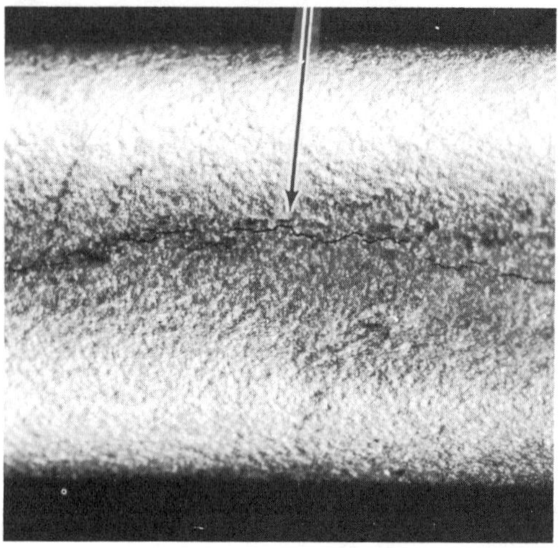

C. HIGHER MAGNIFICATION OF CRACK SHOWING ORANGE PEEL

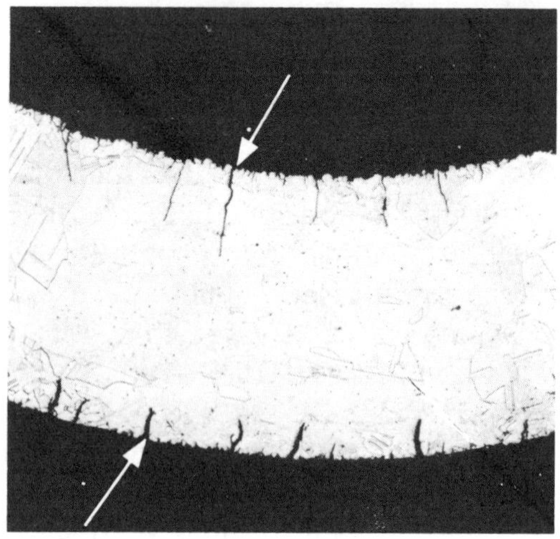

D. MICROGRAPH OF CONVOLUTION WITH PARTIAL CRACKING ON SIDES

Figure 7-10. Convolution Crack Discontinuities

(5) The type of marking material (e.g., grease pencil on titanium) used to identify the area of discontinuities may affect the structure of the article.

b. Ultrasonic Testing Method. Not normally used for the detection of convolution cracks. The configuration of the article (double-walled convolutions) and the prescence of internal micro fractures are all factors that restrict the use of ultrasonics.

c. Eddy Current Testing Method. Not normally used for the detection of convolution cracks. As in the case of ultrasonic testing, the configuration does not lend itself to this method of testing.

d. Liquid Penetrant Testing Method. Not recommended for the detection of convolution cracks. Although the discontinuities are surface, they are internal and are superimposed over an exterior shell which creates a serious problem of entrapment.

e. Magnetic Testing Method. Not applicable. Material is nonferrous.

711 HEAT-AFFECTED ZONE CRACKING

1. Category. Processing (Weldments)

2. Material. Ferrous and Nonferrous

3. Discontinuity Characteristics

Surface. Often quite deep and very tight. Usually run parallel with the weld in the heat-affected zone of the weldment. (See Figure 7-11.)

4. Metallurgical Analysis

Hot cracking of heat-affected zones of weldments increases in severity with increasing carbon content. Steels that contain more than 0.30% carbon are prone to this type of failure and require preheating prior to welding.

5. NDT Methods Application and Limitations

a. Magnetic Particle Testing Method.

(1) Normally used for ferromagnetic weldments.

(2) Prod burns are very detrimental, especially on highly heat-treated articles. Burns may contribute to structural failure of article.

(3) Demagnetization of highly heat-treated articles can be very difficult due to metallurgical structure.

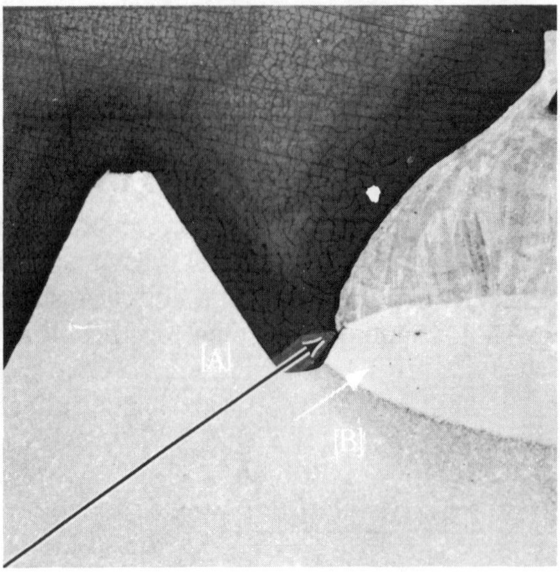

A. MICROGRAPH OF WELD AND HEAT-AFFECTED ZONE SHOWING CRACK. NOTE COLD LAP MASKING THE ENTRANCE OF THE CRACK

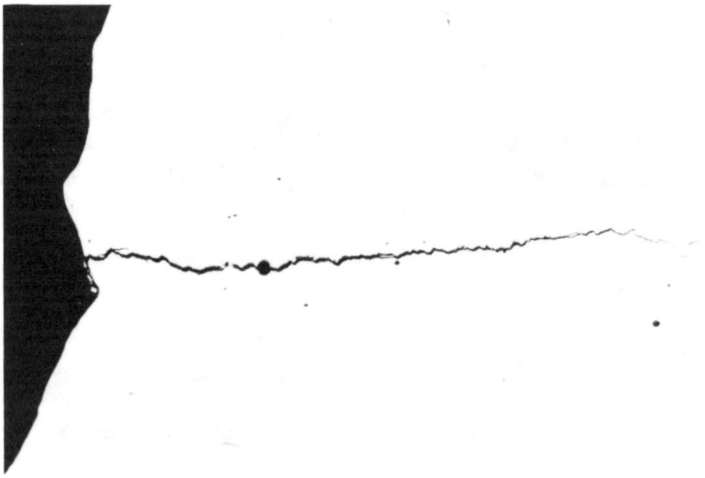

B. MICROGRAPH OF CRACK SHOWN IN (A)

Figure 7-11. Heat-Affected Zone Cracking Discontinuity

b. Liquid Penetrant Testing Method.

(1) Normally used for nonferrous weldments.

(2) Material that has had its surface obliterated, blurred, or blended due to manufacturing processes should not be penetrant tested until the smeared surface has been removed.

(3) Liquid penetrant testing after the application of certain types of chemical film coatings may be invalid due to the covering or filling of the discontinuities.

c. Radiographic Testing Method. Not normally used for the detection of heat-affected zone cracking. Discontinuity orientation and surface origin make other NDT methods more suitable.

d. Ultrasonic Testing Method.

(1) Used where specialized applications have been developed.

(2) Rigid standards and procedures are required to develop valid tests.

(3) The configuration of the surface roughness (i.e., sharp versus rounded root radii and the slope condition) are major factors in deflecting the sound beam.

e. Eddy Current Testing Method. Although not normally used for the detection of heat-affected zone cracking, eddy current testing equipment has the capability of detecting nonferrous surface discontinuities.

712 HEAT-TREAT CRACKS

1. Category. Processing

2. Material . Ferrous and Nonferrous Wrought and Cast Material

3. Discontinuity Characteristics

Surface. Usually deep and forked. Seldom follow a definite pattern and can be in any direction on the part. Originate in areas with rapid change of material thickness, sharp machining marks, fillets, nicks, and discontinuities that have been exposed to the surface of the material. (See Figure 7-12.)

A. FILLET AND MATERIAL THICKNESS CRACKS (TOP CENTER)
RELIEF RADIUS CRACKING (LOWER LEFT)

B. HEAT-TREAT CRACK DUE TO SHARP MACHINING MARKS

Figure 7-12. Heat-Treat Crack Discontinuities

4. Metallurgical Analysis

During the heating and cooling process, localized stresses may be set up by unequal heating or cooling, restricted movement of the article, or unequal cross-sectional thickness. These stresses may exceed the tensile strength of the material causing it to rupture. Where built-in stress risers occur (keyways or grooves) additional cracks may develop.

5. NDT Methods Application and Limitations

a. Magnetic Particle Testing Method.

(1) For ferromagnetic materials, heat-treat cracks are normally detected by magnetic particle testing.

(2) Indications normally appear as straight, forked, or curved indications.

(3) Likely points of origin are areas that would develop stress risers, such as keyways, fillets, or areas with rapid changes in material thickness.

(4) Metallurgical structure of age-hardenable and heat-treatable stainless steels (17.4, 17.7, and 431) may produce nonrelevant indications.

b. Liquid Penetrant Testing Method.

(1) Liquid penetrant testing is the recommended method for nonferrous materials.

(2) Likely points of origin for heat-treat cracks are the same as those listed for magnetic particle testing.

(3) Materials or articles that will eventually be used in LOX systems must be tested with LOX compatible penetrants.

c. Eddy Current Testing Method. Although not normally used for the detection of heat-treat cracks, eddy current testing equipment has the capability of detecting nonferrous surface discontinuities.

d. Ultrasonic Testing Method. Not normally used for detection of heat-treat cracks. If used, the scope pattern will show a definite indication of a discontinuity. Recommended wave mode would be surface.

e. Radiographic Testing Method. Not normally used for detection of heat-treat cracks. Surface discontinuities are more easily detected by other NDT methods designed for surface application.

713 SURFACE SHRINK CRACKS

1. Category. Processing (Welding)

2. Material. Ferrous and Nonferrous

3. Discontinuity Characteristics

Surface. Situated on the face of the weld, fusion zone, and base metal. Range in size from very small, tight, and shallow, to open and deep. Cracks may run parallel or transverse to the direction of welding. (See Figure 7-13.)

4. Metallurgical Analysis

Surface shrink cracks are generally the result of improper heat application, either in heating or welding of the article. Heating or cooling in a localized area may set up stresses that exceed the tensile strength of the material causing the material to crack. Restriction of the movement (contraction or expansion) of the material during heating, cooling, or welding may also set up excessive stresses.

5. NDT Methods Application and Limitations

a. Liquid Penetrant Testing Method.

(1) Surface shrink cracks in nonferrous materials are normally detected by use of liquid penetrants.

(2) Liquid penetrant equipment is easily portable and can be used during in-process control for both ferrous and nonferrous weldments.

(3) Assemblies that are joined by bolting, riveting, intermittent welding, or press fittings will retain the penetrant, which will seep out after developing and mask the adjoining surfaces.

(4) When articles are dried in a hot air dryer or by similar means, excessive drying temperature should be avoided to prevent evaporation of penetrant.

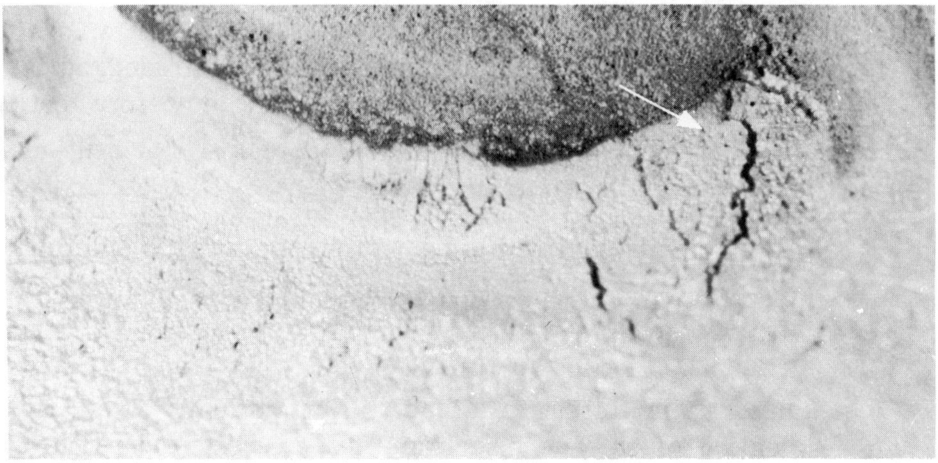

A. TRANSVERSE CRACKS IN HEAT-AFFECTED ZONE

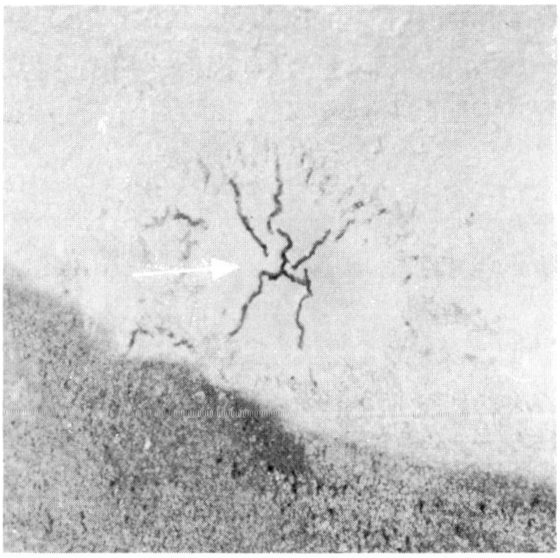

B. TYPICAL STAR-SHAPED CRATER CRACK

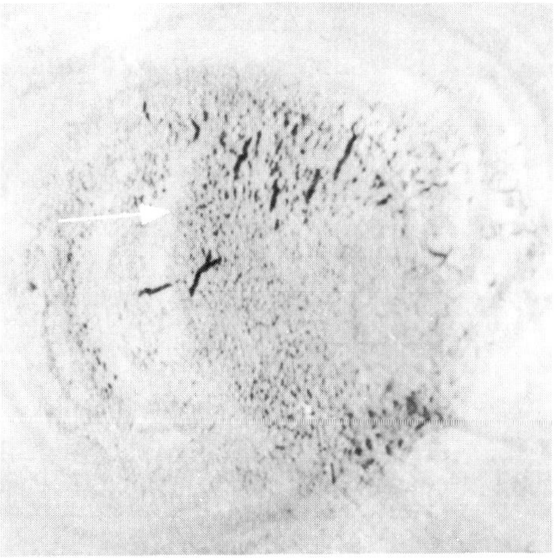

C. SHRINKAGE CRACK AT WELD TERMINAL

Figure 7-13. Surface Shrink Crack Discontinuities

b. Magnetic Particle Testing Method.

(1) Ferromagnetic weldments are normally tested by magnetic particle method.

(2) Surface discontinuities that are parallel to the magnetic field will not produce indications since they do not interrupt or distort the magnetic field.

(3) Areas such as grease fittings, bearing races, or other similar items that might be damaged or clogged by the bath or by the particles should be masked before testing.

c. Eddy Current Testing Method.

(1) Normally confined to nonferrous welded pipe and tubing.

(2) A probe or encircling coil could be used where article configuration permits.

d. Radiographic Testing Method. Not normally used for the detection of surface discontinuities. During the radiographic testing of weldments for other types of discontinuities, surface indications may be detected.

e. Ultrasonic Testing Method. Not normally used for detection of surface shrink cracks. Other forms of NDT (liquid penetrant and magnetic particle) give better results, are more economical, and are faster.

714 THREAD CRACKS

1. Category. Service

2. Material. Ferrous and Nonferrous Wrought Material

3. Discontinuity Characteristics

Surface. Cracks are transverse to the grain (transgranular) starting at the root of the thread. (See Figure 7-14.)

4. Metallurgical Analysis

Fatigue failures of this type are not uncommon. High cyclic stresses resulting from vibration and/or flexing act on the stress risers created by the thread roots to produce cracks. Fatigue cracks may start as fine submicroscopic discontinuities or cracks and propagate in the direction of applied stresses.

A. COMPLETE THREAD ROOT FAILURE

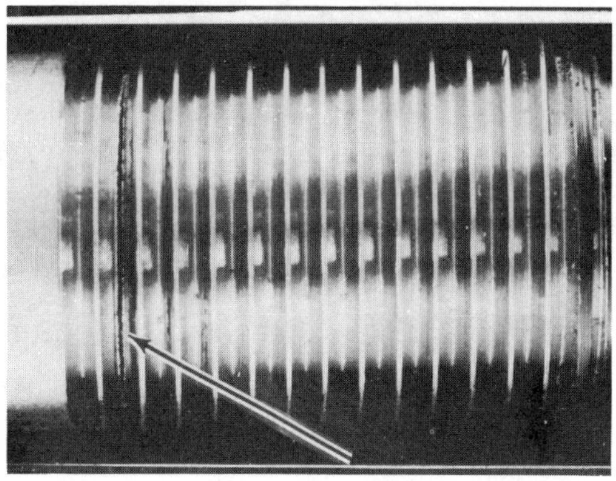

B. TYPICAL THREAD ROOT FAILURE

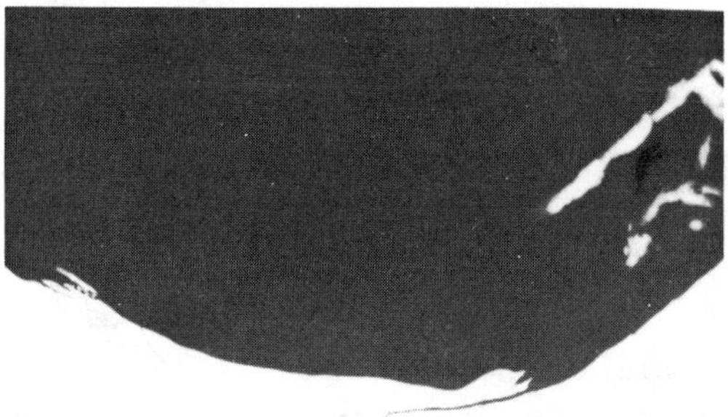

C. MICROGRAPH OF (A) SHOWING CRACK AT BASE OF ROOT

D. MICROGRAPH OF (B) SHOWING TRANS-GRANULAR CRACK AT THREAD ROOT

Figure 7-14. Thread Crack Discontinuities

5. NDT Methods Application and Limitations

a. Liquid Penetrant Testing Method.

(1) Fluorescent penetrant is recommended over nonfluorescent.

(2) Low surface tension solvents such as gasoline and kerosene are not recommended cleaners.

(3) When applying liquid penetrant to components within an assembly or structure, the adjacent areas should be effectively masked to prevent overspraying.

b. Magnetic Particle Testing Method.

(1) Normally used to detect cracks at the threads on ferromagnetic materials.

(2) Nonrelevant magnetic indications may result from the thread configuration.

(3) Cleaning titanium and 440C stainless in halogeneated hydrocarbons may result in structural damage to the material.

c. Eddy Current Testing Method. Not normally used for detecting thread cracks. The article configuration would require specialized equipment if adaptable.

d. Ultrasonic Testing Method. Not recommended for detecting thread cracks. Thread configuration does not lend itself to ultrasonic testing.

e. Radiographic Testing Method. Not recommended for detecting thread cracks. Surface discontinuities are best screened by NDT method designed for the specific condition. Fatigue cracks of this type are very tight and surface connected. Detection by radiography would be extremely difficult.

715 TUBING CRACKS

1. Category. Inherent

2. Material. Nonferrous

3. Discontinuity Characteristics

Tubing cracks formed on the inner surface (I.D.), parallel to direction of grain flow. (See Figure 7-15.)

4. Metallurgical Analysis

Tubing I.D. cracks may be attributed to one or a combination of the following:

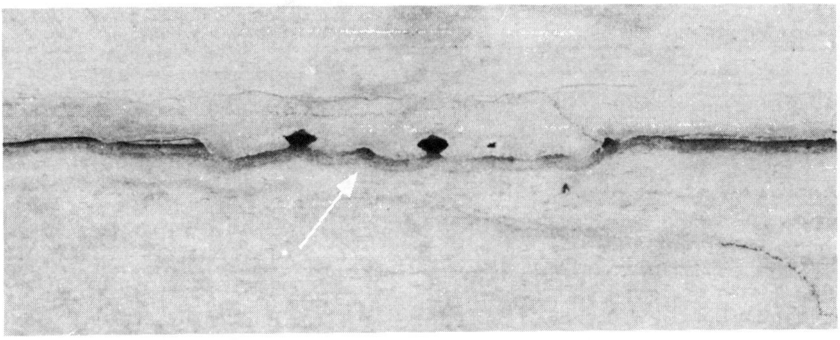

A. TYPICAL CRACK ON INSIDE OF TUBING SHOWING COLD LAP

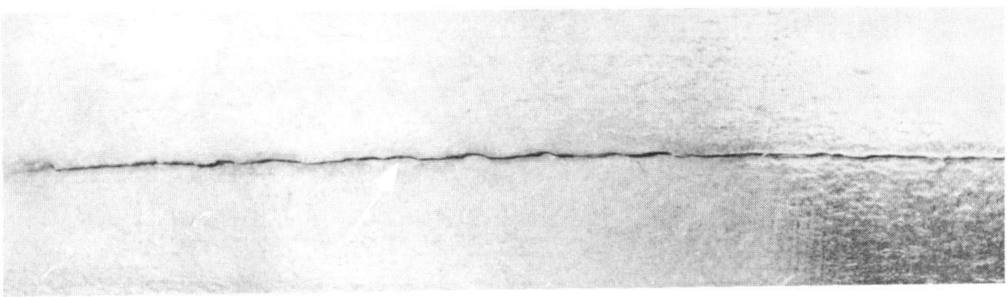

B. ANOTHER PORTION OF SAME CRACK SHOWING CLEAN FRACTURE

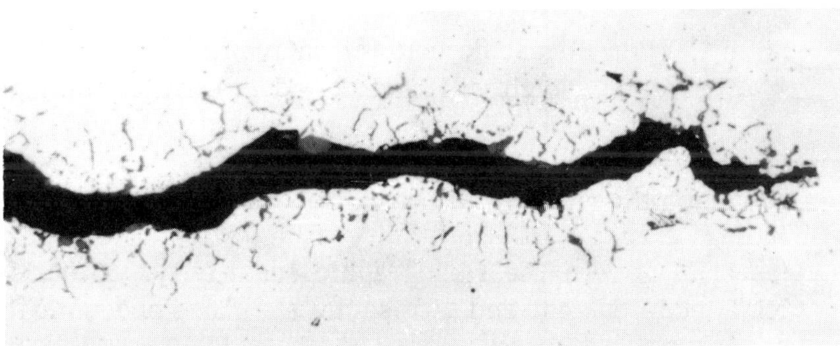

C. MICROGRAPH OF (B)

Figure 7-15. Tubing Crack Discontinuity

a. Improper cold reduction of the tube during fabrication.

b. Foreign material may have been embedded on the inner surface of the tubes causing embrittlement and cracking when the cold worked material was heated during the annealing operation.

c. Insufficient heating rate to the annealing temperature with possible cracking occurring in the 1200-1400°F (645-760°C) range.

5. <u>NDT Methods Application and Limitations</u>

a. Eddy Current Testing Method.

(1) Normally used for detection of this type of discontinuity.

(2) Tube diameters below 1 inch (2.54 cm) and wall thicknesses less than 0.150 inch (3.8 mm) are well within equipment capability.

(3) Testing of ferromagnetic material may be difficult.

b. Ultrasonic Testing Method.

(1) Normally used on tubing.

(2) A wide variety of equipment and transducers are available for screening tubing for internal discontinuities of this type.

(3) Ultrasonic transducers have varying temperature limitations.

(4) Certain ultrasonic contact couplants may have high sulfur content, which will have an adverse effect on high-nickel alloys.

c. Radiographic Testing Method. Not normally used for detecting tubing cracks. Discontinuity orientation and thickness of material govern the radiographic sensitivity. Other forms of NDT (eddy current and ultrasonic) are more economical, faster, and more reliable.

d. Liquid Penetrant Testing Method. Not recommended for detecting tubing cracks. Internal discontinuity would be difficult to process and interpret.

e. Magnetic Particle Testing Method. Not applicable. Material is nonferrous under normal conditions.

716 HYDROGEN FLAKE

1. Category. Processing

2. Material. Ferrous

3. Discontinuity Characteristics

Internal fissures in a fractured surface, flakes appear as bright silvery areas. On an etched surface they appear as short discontinuities. Sometimes known as chrome checks and hairline cracks when revealed by machining. Flakes are extremely thin and generally align parallel with the grain. They are usually found in heavy steel forgings, billets, and bars. (See Figure 7-16.)

4. Metallurgical Analysis

Flakes are internal fissures attributed to stresses produced by localized transformation and decreased solubility of hydrogen during cooling after hot working. Usually found only in heavy alloy steel forgings.

5. NDT Methods Application and Limitations

a. Ultrasonic Testing Method.

(1) Used extensively for the detection of hydrogen flake.

(2) Material in the wrought condition can be screened successfully using either the immersion or the contact method. The surface condition will determine the method most suited.

(3) On the A-scan presentation, hydrogen flake will appear as hash on the screen or as loss of back reflection.

(4) All foreign materials (loose scale, dirt, oil, grease) should be removed prior to any testing. Surface irregularities such as nicks, gouges, tool marks, and scarfing may cause loss of back reflection.

b. Magnetic Particle Testing Method.

(1) Normally used on finished machined articles.

(2) Flakes appear as short discontinuities and resemble chrome checks or hairline cracks.

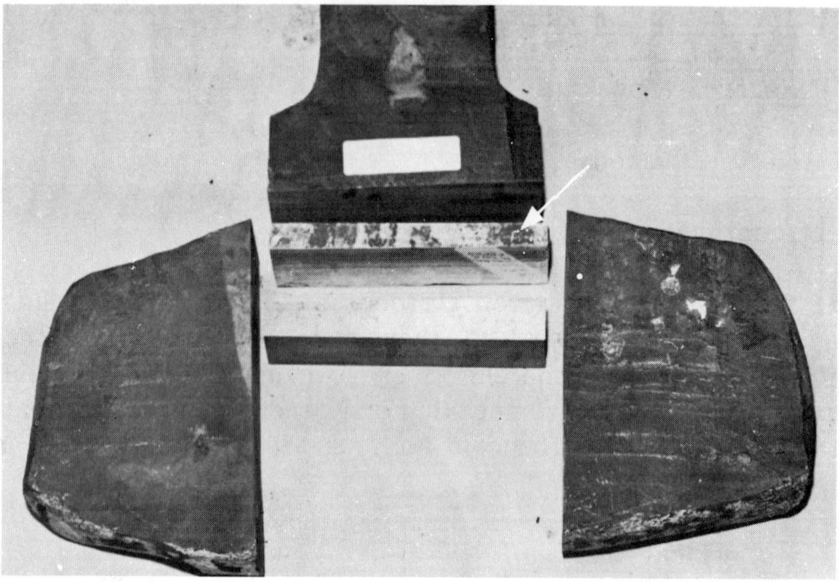

A. 4340 CMS HAND FORGING REJECTED FOR HYDROGEN FLAKE

B. CROSS-SECTION OF (A) SHOWING FLAKE CONDITION IN CENTER OF MATERIAL

Figure 7-16. Hydrogen Flake Discontinuity

(3) Machined surfaces with deep tool marks may obliterate the detection of the flake.

(4) Where the general direction of a discontinuity is questionable, it may be necessary to magnetize in two or more directions.

c. Liquid Penetrant Testing Method. Not normally used for detecting flakes. Discontinuities are very small and tight and would be difficult to detect by liquid penetrants.

d. Eddy Current Testing Method. Not recommended for detecting flakes. The metallurgical structure of ferrous materials limits their adaptability to the use of eddy current.

e. Radiographic Testing Method. Not recommended for detecting flakes. The size of the discontinuity and its location and orientation with respect to the material surface restricts the application of radiography.

717 HYDROGEN EMBRITTLEMENT

1. Category. Processing and Service

2. Material. Ferrous

3. Discontinuity Characteristics

Surface. Small, nondimensional (interface) with no orientation or direction. Found in highly heat-treated material that was subjected to pickling and/or plating or in material exposed to free hydrogen. (See Figure 7-17.)

4. Metallurgical Analysis

Operations such as electroplating or pickling and cleaning prior to electroplating generate hydrogen at the surface of the material. This hydrogen penetrates the surface of the material creating immediate or delayed embrittlement and cracking.

5. NDT Methods Application and Limitations

a. Magnetic Particles Testing Method.

(1) Magnetic indications appear as a fractured pattern.

(2) Hydrogen embrittlement cracks are randomly oriented and may be aligned with the magnetic field.

(3) Magnetic particle testing should be accomplished before and after plating.

A. DETAILED CRACK PATTERN OF HYDROGEN EMBRITTLEMENT

B. HYDROGEN EMBRITTLEMENT UNDER CHROME PLATE

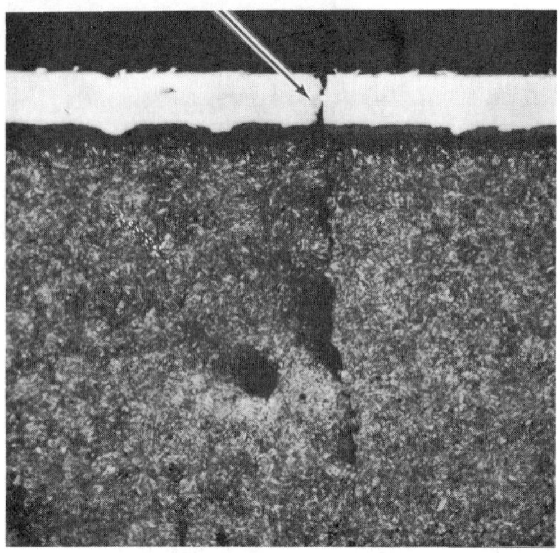

C. HYDROGEN EMBRITTLEMENT PROPAGATED THROUGH CHROME PLATE

Figure 7-17. Hydrogen Embrittlement Discontinuity

(4) Care should be taken so as not to produce nonrelevant indications or cause damage to the article by overheating.

(5) 301 corrosion resistant steel is nonmagnetic in the annealed condition, but becomes magnetic with cold working.

b. Liquid Penetrant Testing Method. Not normally used for detecting hydrogen embrittlement. Discontinuitites on the surface are extremely tight, small, and difficult to detect. Subsequent plating deposit may mask the discontinuity.

c. Ultrasonic Testing Method. Not normally used for detecting hydrogen embrittlement. Article configurations and size do not, in general, lend themselves to this method of testing. Equipment has capability of detecting hydrogen embrittlement. Recommend surface wave technique.

d. Eddy Current Testing Method. Not recommended for detecting hydrogen embrittlement. Many variables inherent in the specific material may produce conflicting patterns.

e. Radiographic Testing Method. Not recommended for detecting hydrogen embrittlement. The sensitivity required to detect hydrogen embrittlement is in most cases in excess of radiographic capabilities.

718 INCLUSIONS

1. Category. Processing (Weldments)

2. Material. Ferrous and Nonferrous Welded Material

3. Discontinuity Characteristics

Surface and subsurface. Inclusions may be any shape. They may be metallic or nonmetallic and may appear singly or be linearly distributed or scattered throughout the weldment. (See Figure 7-18.)

4. Metallurgical Analysis

Metallic inclusions are generally particles of metals of different density as compared to the density of the weld or base metal. Nonmetallic inclusions are oxides, sulphides, slag, or other nonmetallic foreign material entrapped in the weld or trapped between the weld metal and the base metal.

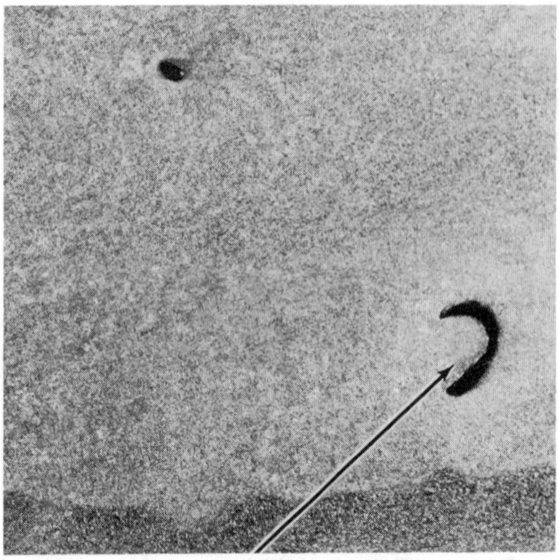

A. METALLIC INCLUSIONS

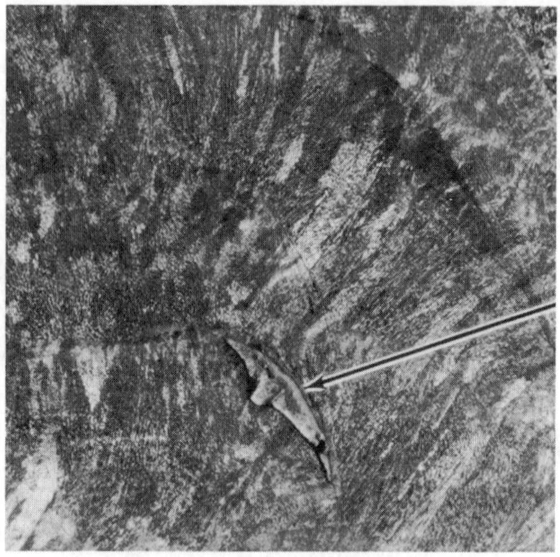

B. INCLUSIONS TRAPPED IN WELD

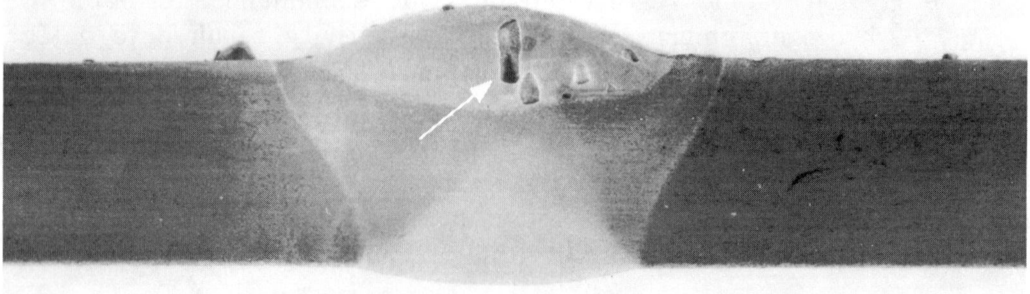

C. CROSS-SECTION OF WELD SHOWING INTERNAL INCLUSIONS

Figure 7-18. Weldment Inclusion Discontinuities

5. NDT Methods Application and Limitations

a. Radiographic Testing Method.

(1) This NDT method is universally used.

(2) Metallic inclusions appear on the radiograph as sharply defined, round, erratically shaped, or elongated white spots and may be isolated or in small linear or scattered groups.

(3) Nonmetallic inclusions will appear on the radiograph as shadows of round globules or elongated or irregularly shaped contours occurring singly, linearly, or scattered throughout the weldment. They will generally appear in

the fusion zone or at the root of the weld. Less absorbent material is indicated by a greater film density and more absorbent materials by a lighter film density.

(4) Foreign material such as loose scales, splatter, or flux may invalidate test results.

b. Eddy Current Testing Method.

(1) Normally confined to thin wall, welded tubing.

(2) Established standards are required if valid results are to be obtained.

c. Magnetic Particle Testing Method.

(1) Normally not used for detecting inclusions in weldments.

(2) Confined to machined weldments where the discontinuities are surface or near surface.

(3) The indications would appear jagged, irregularly shaped, individually or clustered, and would not be too pronounced.

(4) Discontinuities may go undetected when improper contact exists between the magnetic particles and the surface of the article.

d. Ultrasonic Testing Method. Not normally used for detecting inclusions. Specific applications of design or of article configuration, however, may require ultrasonic testing.

e. Liquid Penetrant Testing Method. Not applicable. Inclusions are normally not open fissures.

719 INCLUSIONS

1. Category. Processing

2. Material. Ferrous and Nonferrous Wrought Material

3. Discontinuity Characteristics

Subsurface (original bar) or surface (after machining). There are two types: one is nonmetallic with long straight lines parallel to flow lines and quite tightly adherent. Often short and likely to occur in groups. The other type is nonplastic, appearing as a comparatively large mass not parallel to flow lines. Found in forged, extruded, and rolled material. (See Figure 7-19.)

A. TYPICAL INCLUSION PATTERN ON MACHINED SURFACES

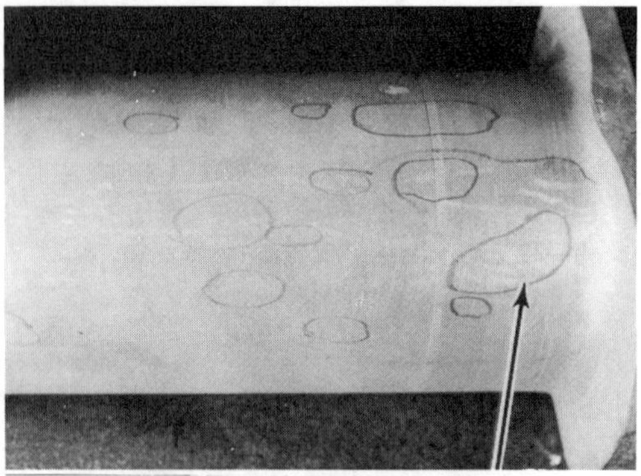

B. STEEL FORGING SHOWING NUMEROUS INCLUSIONS

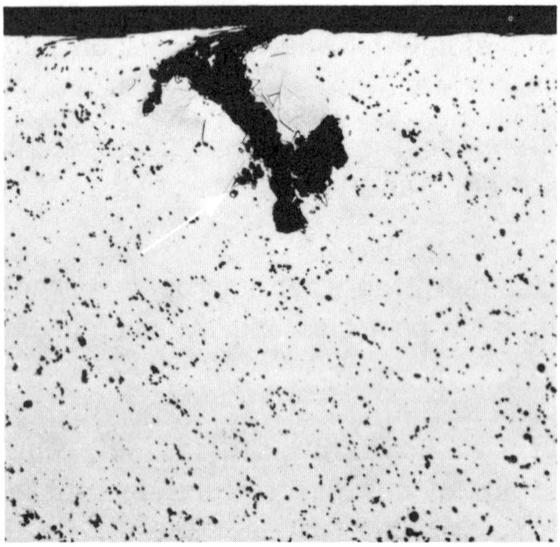

C. MICROGRAPH OF TYPICAL INCLUSION

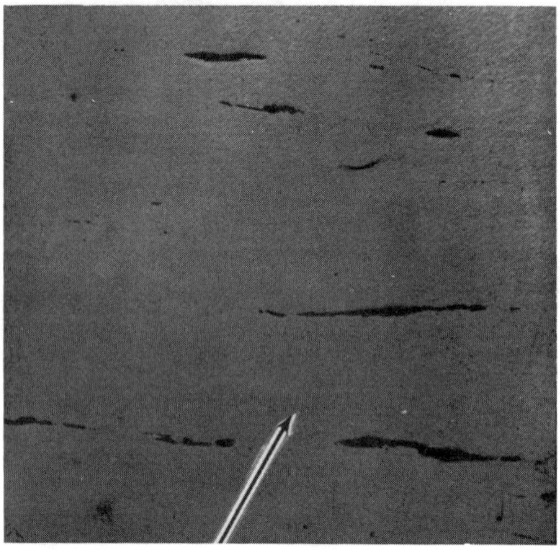

D. LONGITUDINAL CROSS-SECTION SHOWING ORIENTATION OF INCLUSIONS

Figure 7-19. Wrought Inclusion Discontinuities

4. Metallurgical Analysis

Nonmetallic inclusions (stringers) are caused by the existence of slag or oxides in the billet or ingot. Nonplastic inclusions are caused by particles remaining in the solid state during billet melting. Certain types of steels are more prone to inclusions than others.

5. NDT Methods Applications and Limitations

a. Ultrasonic Testing Method.

(1) Normally used to evaluate inclusions in wrought material.

(2) Inclusions will appear as definite interfaces within the metal. Small, clustered condition or conditions on different planes cause a loss in back reflection. Numerous small, scattered conditions cause excessive "noise."

(3) Inclusion orientation in relationship to ultrasonic beam is critical.

(4) The direction of the ultrasonic beam should be perpendicular to the direction of the grain flow whenever possible.

b. Eddy Current Testing Method.

(1) Normally used for thin wall tubing and small diameter rods.

(2) Eddy current testing of ferromagnetic materials can be difficult.

c. Magnetic Particle Testing Method.

(1) Normally used on machined surface.

(2) Inclusions will appear as a straight intermittent or as a continuous indication. They may be individual or clustered.

(3) The magnetizing technique should be such that a surface or near surface inclusion can be satisfactorily detected when its axis is in any direction.

(4) A knowledge of the grain flow of the material is critical since inclusions will be parallel to that direction.

d. Liquid Penetrant Testing Method. Not normally used for detecting inclusions in wrought material. Inclusions are generally not openings in the material surface.

e. Radiographic Testing Method. Not recommended. NDT methods designed for surface testing are more suitable for detecting surface inclusions.

720 LACK OF PENETRATION

1. Category. Processing

2. Material. Ferrous and Nonferrous Weldments

3. Discontinuity Characteristics

Internal or external. Generally irregular and filamentary occurring at the root and running parallel with the weld. (See Figure 7-20.)

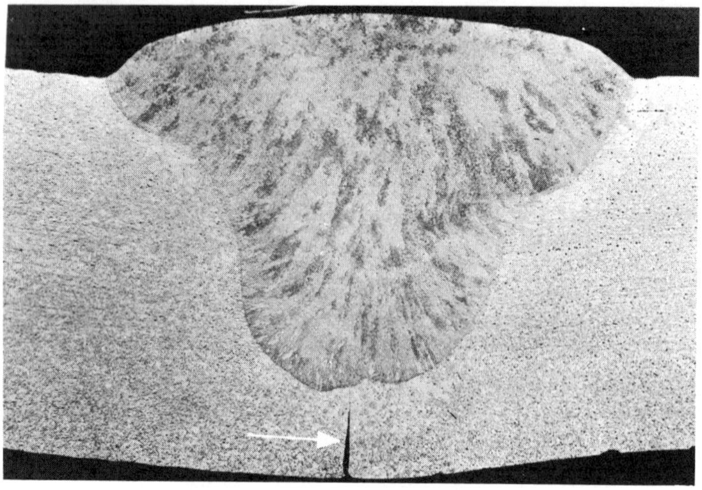

A. INADEQUATE ROOT PENETRATION

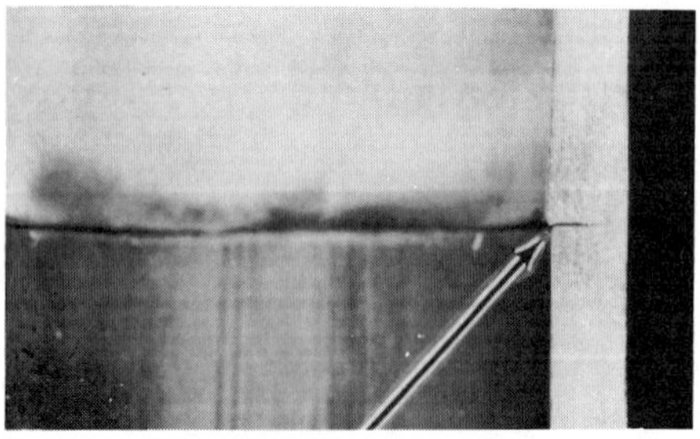

B. INADEQUATE ROOT PENETRATION OF BUTT WELDED TUBE

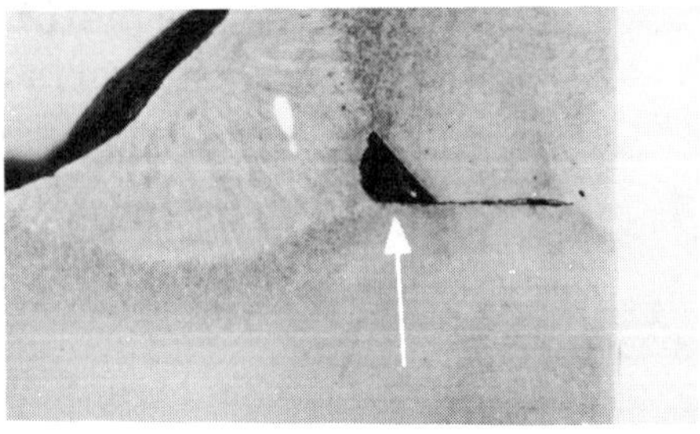

C. INADEQUATE FILLET WELD PENETRATION KNOWN AS BRIDGING

Figure 7-20. Lack of Penetration Discontinuities

4. Metallurgical Analysis

Caused by root face of joint not reaching fusion temperature before weld metal was deposited. Also caused by fast welding rate, too large a welding rod, or too cold a bead.

5. NDT Methods Application and Limitations

a. Radiographic Testing Method.

(1) Used extensively on a wide variety of welded articles to determine the lack of penetration.

(2) Lack of penetration will appear on the radiograph as an elongated, dark area of varying length and width. Lack of penetration may be continuous or intermittent and may appear in the center of the weld at the junction of multipass bends.

(3) Lack of penetration orientation in relationship to the radiographic source is critical.

(4) Sensitivity levels govern the capability to detect small or tight discontinuities.

b. Ultrasonic Testing Method.

(1) Commonly used for specific applications.

(2) Weldments make ultrasonic testing difficult.

(3) Lack of penetration will appear on the scope as a definite break or discontinuity resembling a crack and will give a very sharp reflection.

c. Eddy Current Testing Method.

(1) Normally used to determine lack of penetration in nonferrous welded pipe and tubing.

(2) Eddy current testing can be used where other nonferrous articles can meet the configuration requirement of the equipment.

d. Magnetic Particle Testing Method.

(1) Normally used where backside of weld is visible.

(2) Lack of penetration appears as an irregular indication of varying width.

e. Liquid Penetrant Testing Method.

(1) Normally used where backside of weld is visible.

(2) Lack of penetration appears as an irregular indication of varying width.

(3) Residue left by the penetrant and the developer could contaminate any rewelding operation.

721 LAMINATIONS

1. Category. Inherent

2. Material. Ferrous and Nonferrous Wrought Material

3. Discontinuity Characteristics

Surface and internal. Flat, extremely thin, generally aligned parallel to the work surface of the material. May contain a thin film of oxide between the surfaces. Found in forged, extruded, and rolled material. (See Figure 7-21.)

4. Metallurgical Analysis

Laminations are separations or weaknesses generally aligned parallel to the work surface of the material. They may be the result of pipe, blister, seam, inclusions, or segregations elongated and made directional by working. Laminations are flattened impurities that are extremely thin.

5. NDT Methods Application and Limitations

a. Ultrasonic Testing Method.

(1) For heavier gauge material the geometry and orientation of lamination (normal to the beam) makes their detection limited to ultrasonic testing.

(2) Numerous wave modes may be used depending upon the material thickness or method selected for testing. Automatic and manual contact or immersion methods are adaptable.

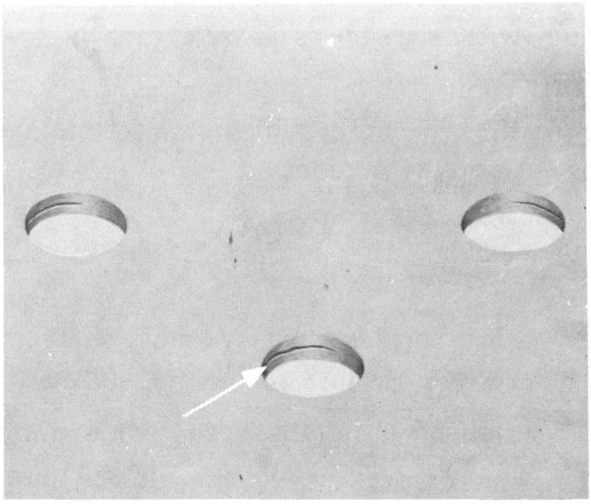

A. LAMINATION IN 0.25 IN. (6.35mm) PLATE

B. LAMINATION IN TITANIUM SHEET

C. LAMINATION IN PLATE SHOWING SURFACE ORIENTATION

D. LAMINATION IN 1 IN. (25.4mm) BAR SHOWING SURFACE ORIENTATION

Figure 7-21. Lamination Discontinuities

(3) Laminations appear as a definite interface with a loss of back reflection.

(4) Through transmission and reflection techniques are applicable for very thin sections.

b. Magnetic Particle Testing Method.

(1) Articles fabricated from ferromagnetic materials are normally tested for lamination by magnetic particle testing methods.

(2) Magnetic indication will appear as a straight, intermittent indication.

(3) Magnetic particle testing is not capable of determining the overall size or depth of the lamination.

c. Liquid Penetrant Testing Method.

(1) Normally used on nonferrous materials.

(2) Machining, honing, lapping, or blasting may smear surface of material and thereby close or mask surface lamination.

(3) Acid and alkalines seriously limit the effectiveness of liquid penetrant testing. Thorough cleaning of the surface is essential.

d. Eddy Current Testing Method. Not normally used to detect laminations. If used, the method must be confined to thin sheet stock.

e. Radiographic Testing Method. Not recommended for detecting laminations. Laminations have very small thickness changes in the direction of the X-ray beam, thereby making radiographic detection almost impossible.

722 LAPS AND SEAMS

1. Category. Processing

2. Material. Ferrous and Nonferrous Rolled Threads

3. Discontinuity Characteristics

Surface. Wavy lines, often quite deep and sometimes very tight, appearing as hairline cracks. Found in rolled threads in the minor pitch, and major diameter of the thread, and in direction of rolling. (See Figure 7-22.)

4. Metallurgical Analysis

During the rolling operation, faulty or oversized dies, or an overfill of material may cause material to be folded over and flattened into the surface of the thread but not fused.

A. TYPICAL AREAS OF FAILURE LAPS AND SEAMS

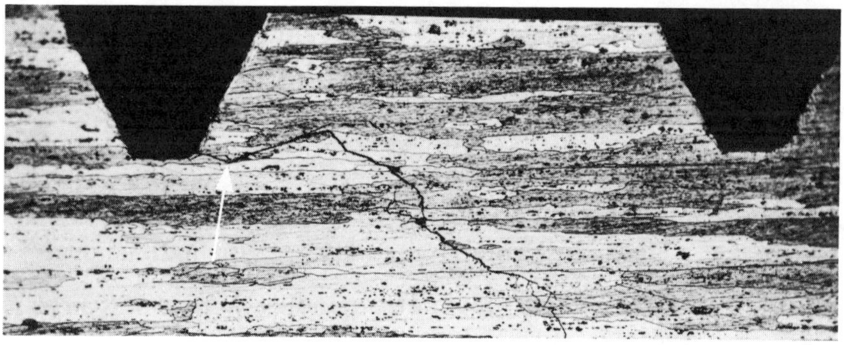

B. FAILURE OCCURRING AT ROOT OF THREAD

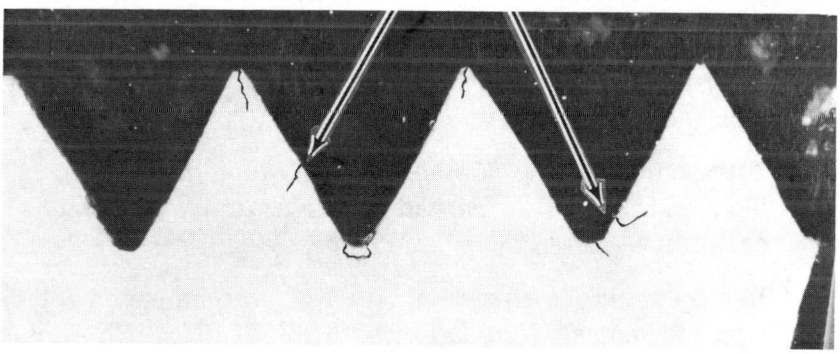

C. AREAS WHERE LAPS AND SEAMS USUALLY OCCUR

Figure 7-22. Lap and Seam Discontinuities in Rolled Threads

5. <u>NDT Methods Application and Limitations</u>

a. Liquid Penetrant Testing Method.

(1) Compatibility with both ferrous and nonferrous materials makes fluorescent liquid penetrant the first choice.

(2) Liquid penetrant indications will be circumferential, slightly curved, intermittent or continuous indications. Laps and seams may occur individually or in clusters.

(3) Foreign material may not only interfere with the penetration of the penetrant into the discontinuity but may cause an accumulation of penetrant in a nondefective area.

(4) Surface of threads may be smeared due to rolling operation, thereby sealing off laps and seams.

(5) Fluorescent and dye penetrants are not compatible. Dye penetrants tend to kill the fluorescent qualities in fluorescent penetrants.

b. Magnetic Particle Testing Method.

(1) Magnetic particle indications of laps and seams generally appear the same as liquid penetrant indications.

(2) Nonrelevant magnetic indications may result from threads.

(3) Questionable magnetic particle indications can be verified by liquid penetrant testing.

c. Eddy Current Testing Method. Not normally used for detecting laps and seams. Article configuration is the restricting factor.

d. Ultrasonic Testing Method. Not recommended for detecting laps and seams. Thread configurations restrict ultrasonic capability.

e. Radiographic Testing Method. Not recommended for detecting laps and seams. Size and orientation of discontinuities restricts the capability of radiographic testing.

723 LAPS AND SEAMS

1. Category. Processing

2. Material. Ferrous and Nonferrous Wrought Material

3. Discontinuity Characteristics

a. Lap Surface. Wavy lines - usually not very pronounced or tightly adherent since they usually enter the surface at a small angle. Laps may have surface openings smeared closed. Found in wrought forgings, plate, tubing, bar, and rod. (See Figure 7-23.)

A. TYPICAL FORGING LAP

B. MICROGRAPH OF A LAP

Figure 7-23. Lap and Seam Discontinuities in Wrought Material

b. Seam Surface. Lengthy, often quite deep and sometimes very tight; usually occur in parallel fissures with the grain; and, at times, spiral when associated with rolled rod and tubing.

4. Metallurgical Analysis

Seams originate from blowholes, cracks, splits, and tears introduced in earlier processing and elongated in the direction of rolling or forging. The distance between adjacent innerfaces of the discontinuity is very small.

Laps are similar to seams and may result from improper rolling, forging, or sizing operations. During the processing of the material, corners may be folded over or an overfill may exist during sizing that results in material being flattened, but not fused into the surface. Laps may occur on any part of the article.

5. NDT Methods Application and Limitations

a. Magnetic Particle Testing Method.

(1) Magnetic particle testing is recommended for ferromagnetic material.

(2) Surface and near-surface laps and seams may be detected by this method.

(3) Laps and seams may appear as straight, spiral, or slightly curved indications. They may be individual or clustered and continuous or intermittent.

(4) Magnetic buildup at laps and seams is very small. Therefore a magnetizing current greater than that used for the detection of cracks is necessary.

(5) Correct magnetizing technique should be used when examining for forging laps since the discontinuity may lie in a plane nearly parallel to the surface.

b. Liquid Penetrant Testing Method.

(1) Liquid penetrant testing is recommended for nonferrous material.

(2) Laps and seams may be very tight and difficult to detect especially by liquid penetrant.

(3) Liquid penetrant testing of laps and seams can be improved slightly by heating the article before applying the penetrant.

c. Ultrasonic Testing Method.

(1) Normally used to test wrought material prior to machining.

(2) Surface wave technique permits accurate evaluation of the depth, length, and size of laps and seams.

(3) Ultrasonic indications of laps and seams will appear as definite inner faces within the metal.

d. Eddy Current Testing Method.

(1) Normally used for the evaluation of laps and seams in tubing and pipe.

(2) Other articles can be screened by eddy current where article configuration and size permit.

e. Radiographic Testing Method. Not recommended for detecting laps and seams in wrought material.

724 MICROSHRINKAGE

1. Category. Processing

2. Material. Magnesium Casting

3. Discontinuity Characteristics

Internal. Small filamentary voids in the grain boundaries appear as concentrated porosity in cross section. (See Figure 7-24.)

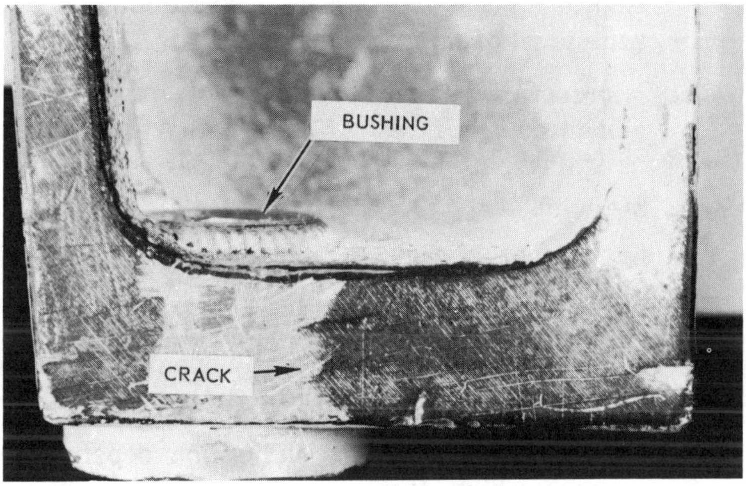

A. CRACKED MAGNESIUM HOUSING

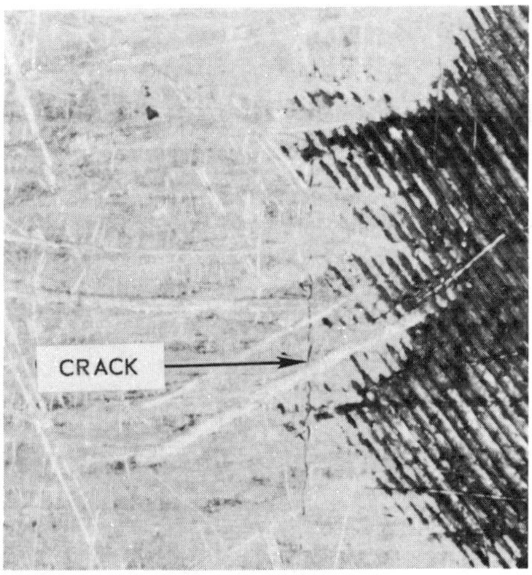

B. CLOSE-UP VIEW OF (A)

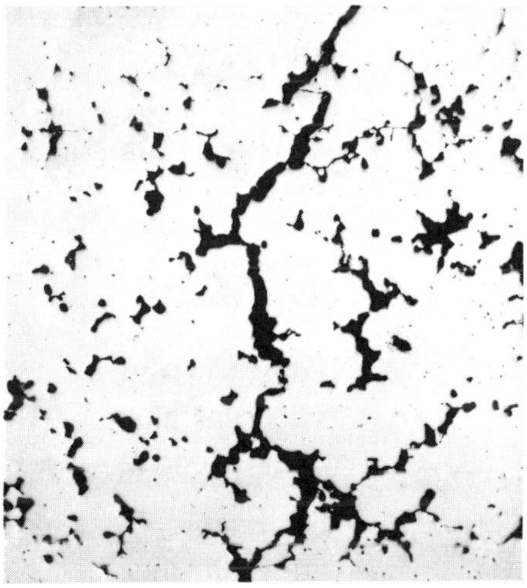

C. MICROGRAPH OF CRACKED AREA

Figure 7-24. Microshrinkage Discontinuity

4. Metallurgical Analysis

Shrinkage occurs while the metal is in a plastic or semimolten state. If sufficient molten metal cannot flow into different areas as it cools, the shrinkage will leave a void. The void is identified by its appearance and by the time in the plastic range it occurs. Microshrinkage is caused by the withdrawal of the low melting point constituent from the grain boundaries.

5. NDT Methods Application and Limitations

a. Radiographic Testing Method.

(1) Radiography is universally used to determine the acceptance level of microshrinkage.

(2) Microshrinkage will appear on the radiograph as an elongated swirl resembling feathery streaks or as dark irregular patches that are indicative of cavities in the grain boundaries.

b. Liquid Penetrant Testing Method.

(1) Normally used on finished machined surfaces.

(2) Microshrinkage is not normally open to the surface. These conditions will, therefore, be detected in machined areas.

(3) The appearance of the indication depends on the plane through which the microshrinkage has been cut. The appearance varies from a continuous hairline to a massive porous indication.

(4) Penetrant may act as a contaminant by saturating the microporous casting affecting its ability to accept a surface treatment.

(5) Serious structural or dimensional damage to the article can result from the improper use of acids or alkalies. They should never be used unless approval is obtained.

c. Eddy Current Testing Method. Not recommended for detecting microshrinkage. Article configuration and type of discontinuity do not lend themselves to eddy current testing.

d. Ultrasonic Testing Method. Not recommended for detecting microshrinkage. Cast structure and article configuration are restricting factors.

e. Magnetic Particle Testing Method. Not applicable. Material is nonferrous.

725 GAS POROSITY

1. Category. Processing

2. Material. Ferrous and Nonferrous Weldments

3. Discontinuity Characteristics

Surface or subsurface. Rounded or elongated, teardrop shaped, with or without a sharp discontinuity at the point. Scattered uniformly throughout the weld or isolated in small groups. May also be concentrated at the root or toe. (See Figure 7-25.)

4. Metallurgical Analysis

Porosity in welds is caused by gas entrapment in the molten metal, too much moisture on the base or filler metal, or improper cleaning or preheating.

5. NDT Methods Application and Limitations

a. Radiography Testing Method.

(1) Radiography is the most universally used NDT method for the detection of gas porosity in weldments.

(2) The radiograhic image of a "round" porosity will appear as oval shaped spots with smooth edges, while "elongated" porosity will appear as oval shaped spots with the major axis sometimes several times longer than the minor axis.

(3) Foreign material such as loose scale, flux, or splatter will affect validity of test results.

b. Ultrasonic Testing Method.

(1) Ultrasonic testing equipment is highly sensitive, capable of detecting microseparations. Established standards should be used if valid test results are to be obtained.

(2) Surface finish and grain size will affect the validity of the test results.

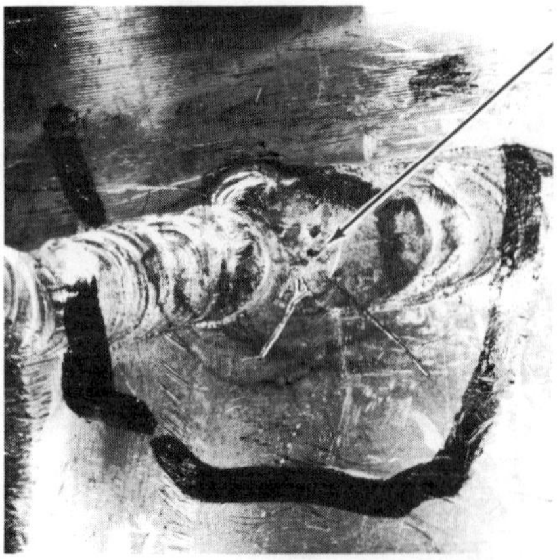

A. TYPICAL SURFACE POROSITY

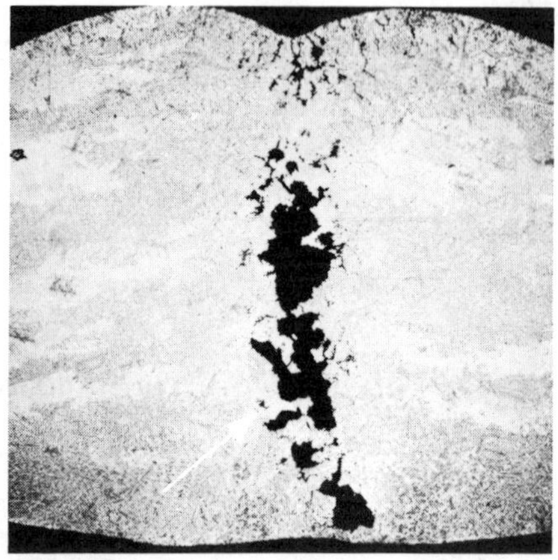

B. CROSS-SECTION OF (A) SHOWING EXTENT OF POROSITY

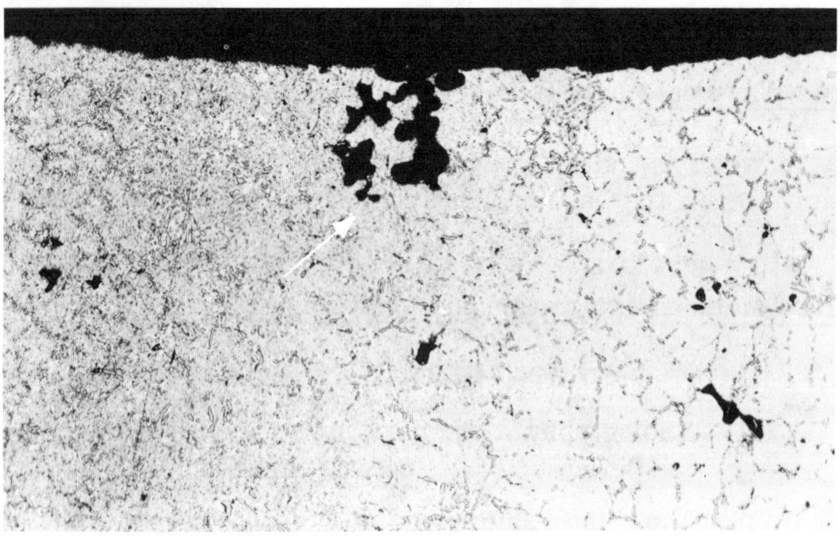

C. MICROGRAPH OF CROSS-SECTION SHOWING TYPICAL SHRINKAGE POROSITY

Figure 7-25. Gas Porosity Discontinuity

c. Eddy Current Testing Method.

(1) Normally confined to thin-wall welded pipe and tube.

(2) Penetration restricts testing to a depth of more than one-quarter inch.

d. Liquid Penetrant Testing Method.

(1) Normally confined to inprocess control of ferrous and nonferrous weldments.

(2) Liquid penetrant testing, like magnetic particle, is restricted to surface evaluation.

(3) Extreme caution must be exercised to prevent any cleaning material, magnetic (iron oxide), and liquid penetrant materials from becoming entrapped and contaminating the rewelding operation.

e. Magnetic Particle Testing Method. Not normally used to detect gas porosity. Only surface porosity would be evident. Near surface porosity would not be clearly defined since indications are neither strong nor pronounced.

726 UNFUSED POROSITY

1. Category. Processing

2. Material. Aluminum

3. Discontinuity Characteristics

Internal. Wafer-thin fissures aligned parallel with the grain flow. Found in wrought aluminum that has been rolled, forged, or extruded. (See Figure 7-26.)

4. Metallurgical Analysis

Unfused porosity is attributed to porosity in the cast ingot. During the rolling, forging, or extruding operations it is flattened into wafer-thin shape. If the internal surface of these discontinuities is oxidized or is composed of a foreign material, they will not fuse during the subsequent processing, which results in an extremely thin interface or void.

5. NDT Methods Application and Limitations

a. Ultrasonic Testing Method.

(1) Used extensively for the detection of unfused porosity.

(2) Raw materials may be tested in the "as received" configuration.

(3) Ultrasonic testing fixes the location of the void in all three directions.

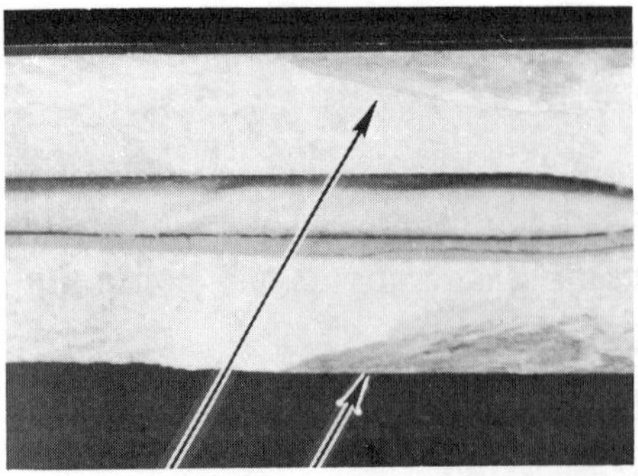

A. FRACTURED SPECIMEN SHOWING UNFUSED POROSITY

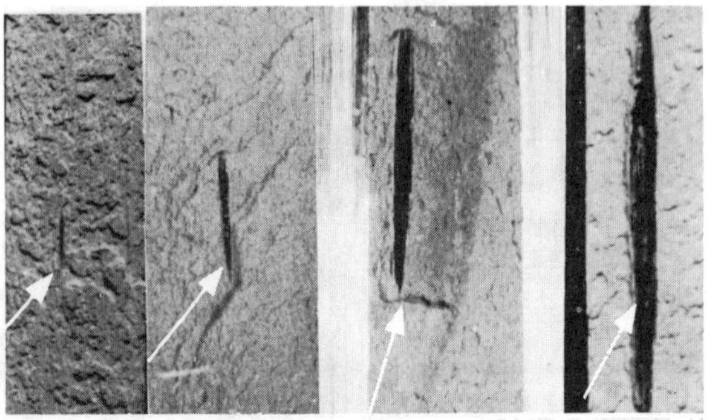

B. UNFUSED POROSITY EQUIVALENT TO 1/64 IN. (0.40 mm), 3/64 IN. (1.17 mm) 5/64 IN. (1.98 mm) AND 8/64 IN. (3.18 mm) (left to right)

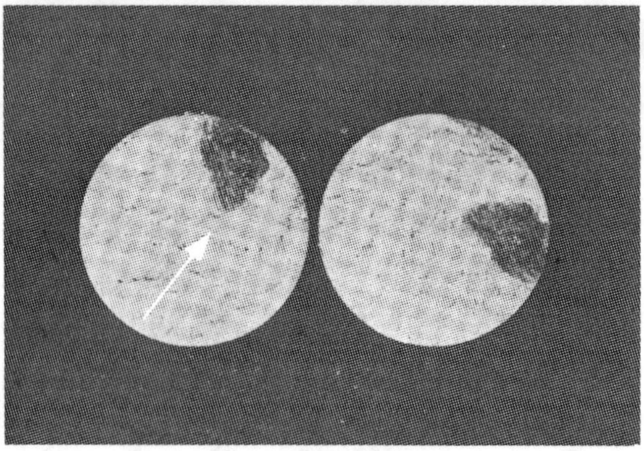

C. TYPICAL UNFUSED POROSITY

Figure 7-26. Unfused Porosity Discontinuity

(4) Where the general direction of the discontinuity is unknown, it may be necessary to test from several directions.

(5) Method of manufacture and subsequent article configuration will determine the orientation of the unfused porosity to the material surface.

b. Liquid Penetrant Testing Method.

(1) Normally used on nonferrous, machined articles.

(2) Unfused porosity will appear as a straight line of varying lengths running parallel with the grain. Liquid penetrant testing is restricted to surface evaluation.

(3) Surface preparations such as vapor blasting, honing, grinding, or sanding may obliterate possible indications by

masking the surface discontinuities, thereby restricting the reliability of liquid penetrant testing.

(4) Excessive agitation of penetrant materials may produce foaming.

c. Eddy Current Testing Method. Not normally used for detecting unfused porosity.

d. Radiographic Testing Method. Not normally used for detecting unfused porosity. Wafer-thin discontinuities are difficult to detect by a method that measures density or that requires that the discontinuity be perpendicular to the X-ray beam.

e. Magnetic Particle Testing Method. Not applicable. Material is nonferrous.

727 STRESS CORROSION

1. Category. Service

2. Material. Ferrous and Nonferrous

3. Discontinuity Characteristics

Surface. Range from shallow to very deep, and usually follow the grain flow of the material; however, transverse cracks are also possible. (See Figure 7-27.)

4. Metallurgical Analysis

Three factors are necessary for the phenomenon of stress corrosion to occur: 1) a sustained static tensile stress, 2) the presence of a corrosive environment, and 3) the use of a material that is susceptible to this type of failure. Stress corrosion is much more likely to occur at high levels of stress than at low levels of stress. The type of stresses include residual (internal) as well as those from external (applied) loading.

5. NDT Methods Application and Limitations

a. Liquid Penetrant Testing Method.

(1) Liquid penetrant is normally used for the detection of stress corrosion.

Figure 7-27. Stress Corrosion Discontinuity

(2) In the preparation, application, and final cleaning of articles, extreme care must be exercised to prevent overspraying and contamination of the surrounding articles.

(3) Chemical cleaning immediately before the application of liquid penetrant may seriously affect the test results if the solvents are not given time to evaporate.

(4) Service articles may contain moisture within the discontinuity which will dilute, contaminate, and invalidate results if the moisture is not removed.

b. Eddy Current Testing Method. Not normally used to detect stress corrosion. Eddy current equipment is capable of resolving stress corrosion where article configuration is compatible with equipment limitations.

c. Ultrasonic Testing Method. Not normally used to detect stress corrosion. Discontinuities are perpendicular to surface of material and require surface technique.

d. Magnetic Particle Testing Method. Not normally used to detect stress corrosion. Configuration of article and usual non-ferromagnetic condition exclude magnetic particle testing.

e. Radiographic Testing Method. Not normally used to detect stress corrosion. Surface indications are best detected by NDT method designed for such application. However, radiography can and has shown stress corrosion with the use of the proper technique.

728 HYDRAULIC TUBING

1. Category. Processing and Service

2. Material. Aluminum 6061-T6

3. Discontinuity Characteristics

Surface and internal. Range in size from short to long, shallow to very tight and deep. Usually they will be found in the direction of the grain flow with the exception of stress corrosion, which has no direction. (See Figure 7-28.)

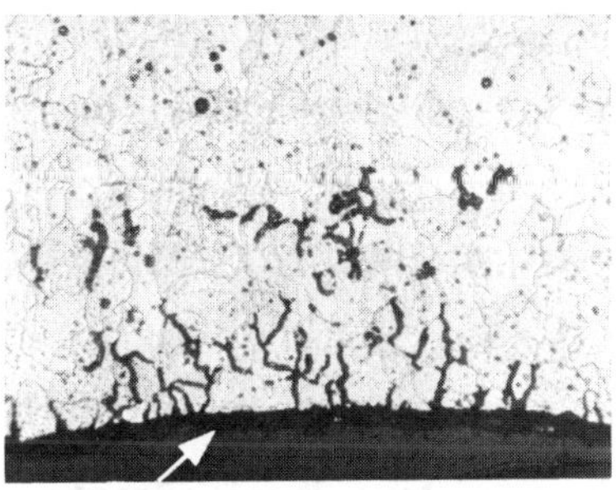

A. INTERGRANULAR CORROSION

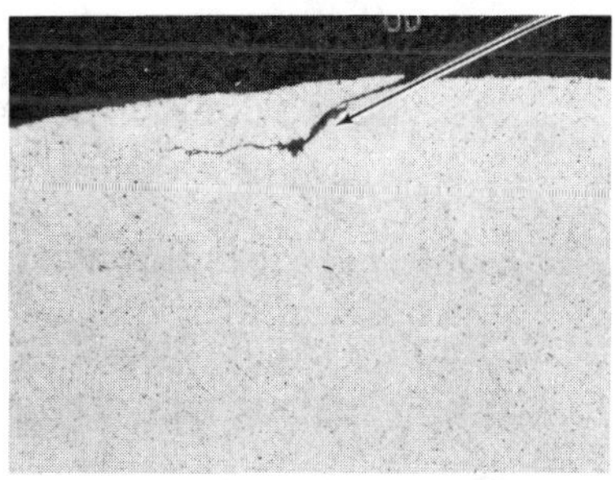

B. LAP IN OUTER SURFACE OF TUBING

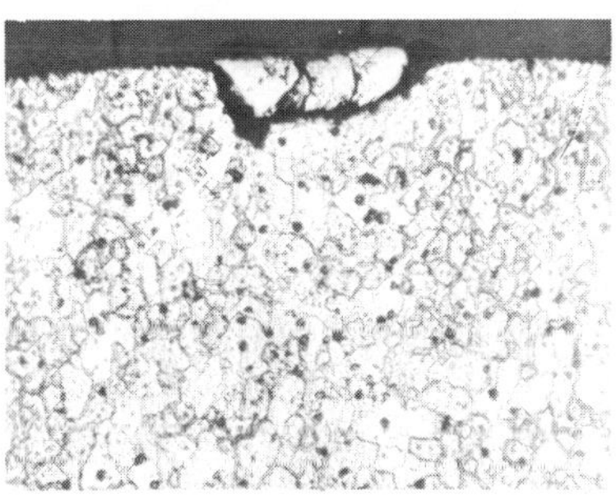

C. EMBEDDED FOREIGN MATERIAL

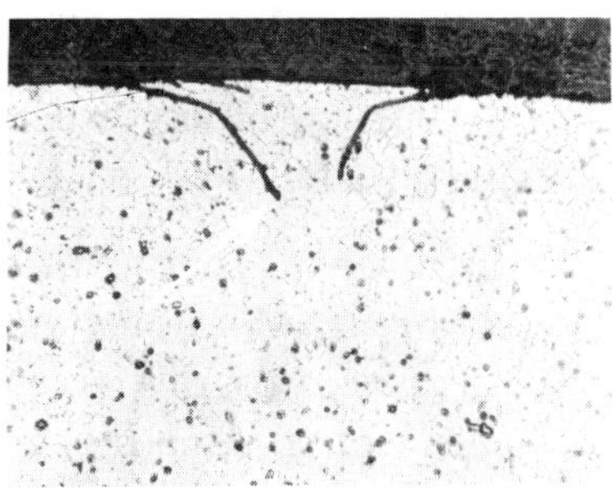

D. TWIN LAPS IN OUTER SURFACE OF TUBING

Figure 7-28. Hydraulic Tubing Discontinuities

4. Metallurgical Analysis

Hydraulic tubing discontinuities are usually one of the following:

a. Foreign material coming in contact with the tube material and being embedded into the surface of the tube.

b. Laps which are the result of material being folded over and not fused.

c. Seams which originate from blowholes, cracks, splits and tears introduced in the earlier processing, and then are elongated during rolling.

d. Intergranular corrosion which is due to the presence of a corrosive environment.

5. NDT Methods Application and Limitations

a. Eddy Current Testing Method.

(1) Universally used for testing of nonferrous tubing.

(2) Heavier-walled tubing, 0.25 in. (6.3 mm) and over, may not be successfully tested due to the penetration ability of the equipment.

(3) The specific nature of various discontinuities may not be clearly defined.

(4) Test results will not be valid unless controlled by known standards.

(5) Testing of ferromagnetic material may be difficult.

(6) All material should be free of any foreign material that would invalidate the test results.

b. Liquid Penetrant Testing Method. Not normally used for detecting tubing discontinuities. Eddy current is more economical, faster, and, with established standards, is more reliable.

c. Ultrasonic Testing Method. Not normally used for detecting tubing discontinuities. Eddy current is recommended over ultrasonic testing since it is faster and more economical for this range of surface discontinuity and nonferrous material.

d. Radiographic Testing Method. Not normally used for detecting tubing discontinuities. The size and type of discontinuity and the configuration of the article limit the use of radiography for screening of material for this group of discontinuities.

e. Magnetic Particle Testing Method. Not applicable. Material is nonferrous.

729 MANDREL DRAG

1. Category. Processing

2. Material. Nonferrous Thick-Wall Seamless Tubing

3. Discontinuity Characteristics

Internal surface of thick-wall tubing. Range from shallow even gouges to ragged tears. Often a slug of the material will be embedded within the gouged area. (See Figure 7-29.)

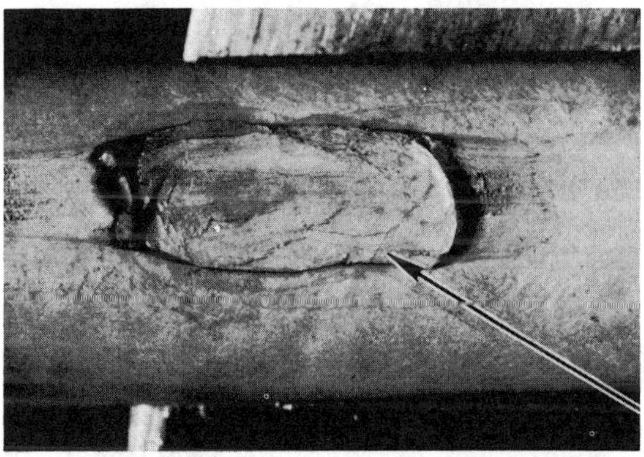

A. EMBEDDED SLUG SHOWING DEEP GOUGE MARKS

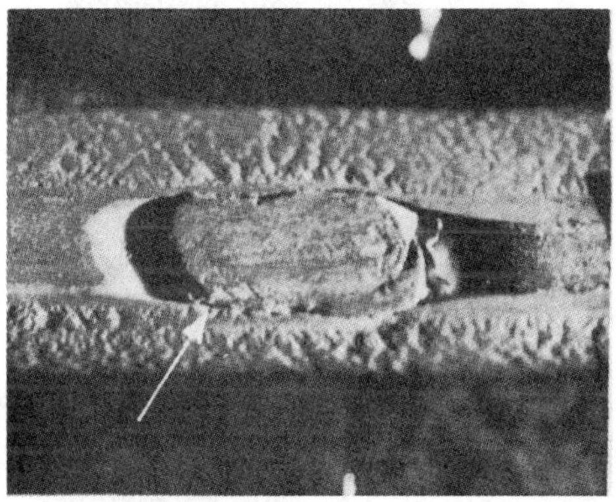

B. SLUG BROKEN LOOSE FROM TUBING WALL

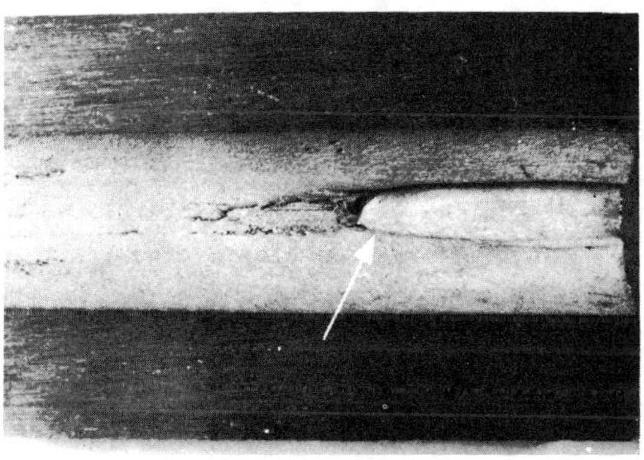

C. ANOTHER TYPE OF EMBEDDED SLUG

D. GOUGE ON INNER SURFACE OF PIPE

Figure 7-29. Mandrel Drag Discontinuities

4. Metallurgical Analysis

During the manufacture of thick-wall seamless tubing, the billet is ruptured as it passes through the offset rolls. As the piercing mandrel follows this fracture, a portion of the material may break loose and be forced over the mandrel. As it does, the surface of the tubing may be scored or have the slug embedded into the wall. Certain types of material are more prone to this type of failure than others.

5. NDT Methods Application and Limitations

a. Eddy Current Testing Method.

(1) Normally used for the testing of thin-wall pipe or tube.

(2) Eddy current testing may be confined to nonferrous materials.

(3) Discontinuities are qualitative, not quantative indications.

(4) Several factors simultaneously affect output indications.

b. Ultrasonic Testing Method.

(1) Normally used for the screening of thick-wall pipe or tube for mandrel drag.

(2) Can be used to test both ferrous and nonferrous pipe or tube.

(3) May be used in support of production line since it is adaptable for automatic instrumentation.

(4) Configuration of mandrel drag or tear will produce very sharp and noticeable indications on the scope.

c. Radiographic Testing Method. Not normally used although it has been instrumental in the detection of mandrel drag during examination of adjacent welds. Complete coverage requires several exposures around the circumference of the tube. This method is not designed for production support since it is very slow and costly for large volumes of pipe or tube. Radiograph will disclose only two dimensions and not the third.

d. Liquid Penetrant Testing Method. Not recommended for detecting mandrel drag since discontinuity is internal and would not be detectable.

e. Magnetic Particle Testing Method. Not recommended for detecting mandrel drag. Discontinuities are not close enough to the surface to be detectable by magnetic particles. Most mandrel drag will occur in seamless stainless steel.

730 SEMICONDUCTORS

1. Category. Processing and Service

2. Material. Hardware

3. Discontinuity Characteristics

Internal. Appear in many sizes and shapes and various degrees of density. They may be misformed, misaligned, damaged, or may have broken internal hardware. Found in transistors, diodes, resistors, and capacitors. (See Figure 7-30.)

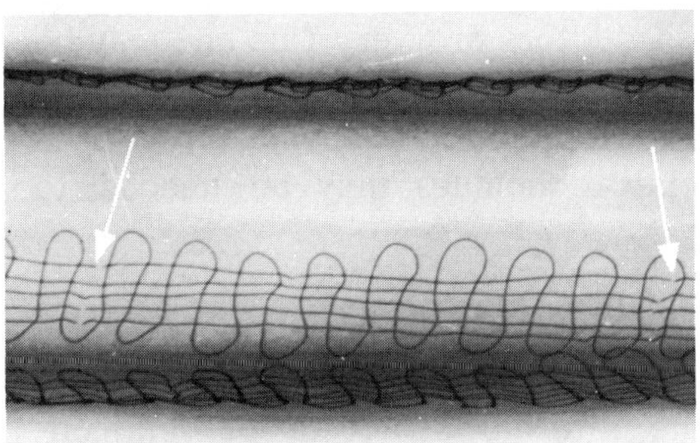

A. STRANDS BROKEN IN HEATER BLANKET

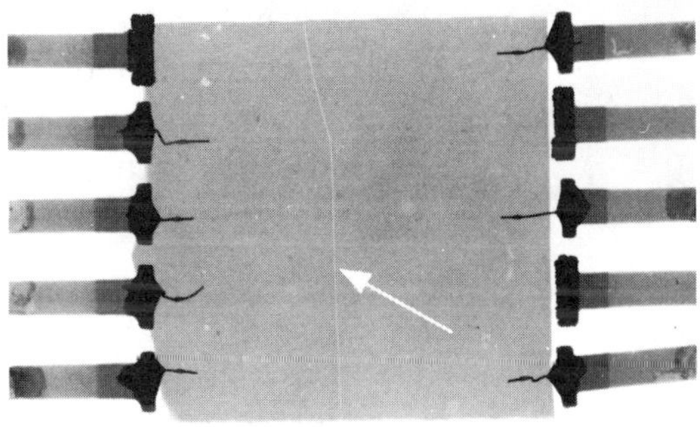

B. FINE CRACK IN PLASTIC CASING MATERIAL

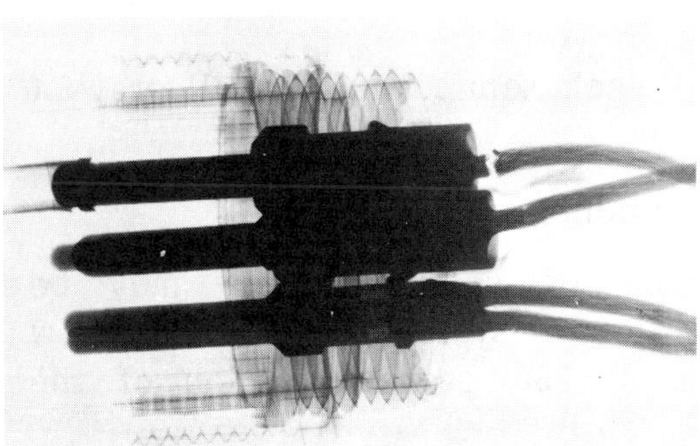

C. BROKEN ELECTRICAL CABLE

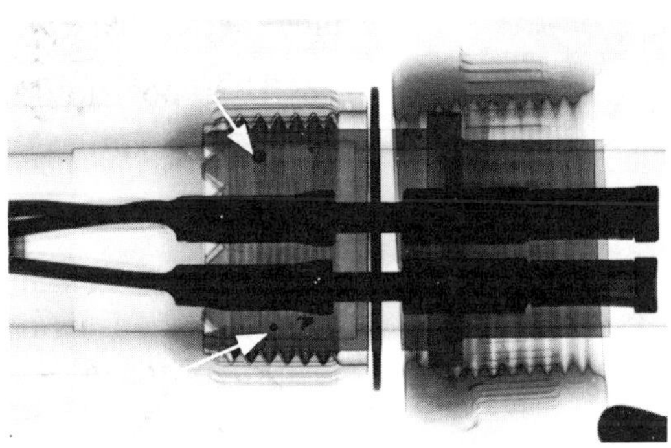

D. FOREIGN MATERIAL WITHIN SEMICONDUCTOR

Figure 7-30. Semiconductor Discontinuities

4. Metallurgical Analysis

Semiconductor discontinuities such as loose wire, weld splash, flakes, solder balls, loose leads, inadequate clearance between internal elements and case, and inclusions or voids in seals or around lead connections are the product of processing errors.

5. NDT Methods Application and Limitations

a. Radiographic Testing Method.

(1) Universally used as the NDT method for the detection of discontinuities in semiconductors.

(2) The configuration and internal structure of the various semiconductors limit the NDT method to radiography.

(3) Semiconductors that have copper heat sinks may require more than one technique due to the density of the copper.

(4) Internal wires in semiconductors are very fine and may be constructed from materials of different density such as copper, silver, gold and aluminum. If the latter is used with the others, special techniques may be needed to resolve test reliability.

(5) Microparticles may require the highest sensitivity to resolve.

(6) The complexity of the internal structure of semiconductors may require additional views to exclude the possibility of non-detection of discontinuities due to masking by hardware.

(7) Positive positioning of each semiconductor will prevent invalid interpretation.

(8) Source angle should give minimum distortion.

(9) Preliminary examination of semiconductors may be accomplished using a vidicon system that would allow visual observation during 360 degree rotation of the article.

b. Eddy Current Testing Method. Not recommended for detecting semiconductor discontinuities. Nature of discontinuity and method of construction of the article do not lend themselves to this form of NDT.

c. Magnetic Particle Testing Method. Not recommended for detecting semiconductor discontinuities.

d. Liquid Penetrant Testing Method. Not recommended for detecting semiconductor discontinuities.

e. Ultrasonic Testing Method. Not recommended for detecting semiconductor discontinuities.

731 HOT TEARS

1. Category. Inherent

2. Material. Ferrous Castings

3. Discontinuity Characteristics

Internal or near surface. Appear as ragged line of variable width and numerous branches. Occur singly or in groups. (See Figure 7-31.)

4. Metallurgical Analysis

Hot cracks (tears) are caused by nonuniform cooling resulting in stresses which rupture the surface of the metal while its temperature is still in the brittle range. Tears may originate where stresses are set up by the more rapid cooling of thin sections that adjoin heavier masses of metal, which are slower to cool.

5. NDT Methods Application and Limitations

a. Radiographic Testing Method.

(1) Radiographic testing is the first choice since the material is cast structure and the discontinuities may be internal and surface.

(2) Orientation of the hot tear in relation to the source may influence the test results.

(3) The sensitivity level may not be sufficient to detect fine surface hot tears.

b. Magnetic Particle Testing Method.

(1) Hot tears that are exposed to the surface can be screened with magnetic particle method.

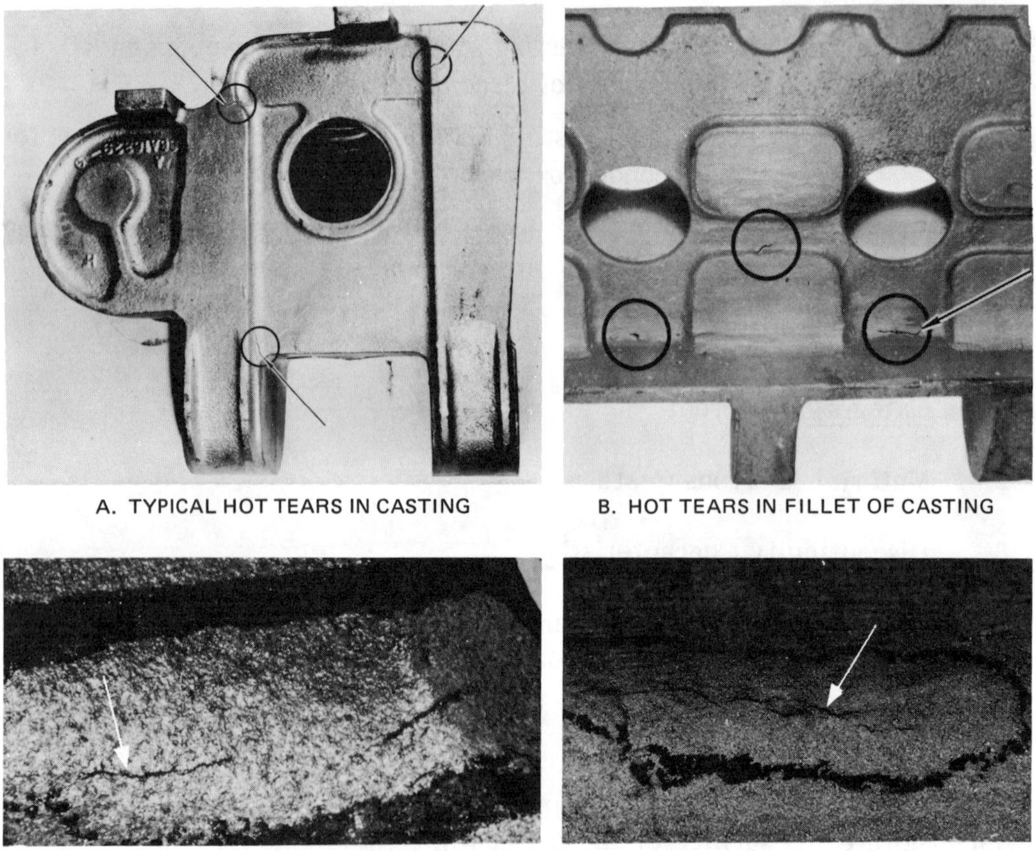

A. TYPICAL HOT TEARS IN CASTING

B. HOT TEARS IN FILLET OF CASTING

C. CLOSE-UP OF HOT TEARS IN (A)

D. CLOSE-UP OF HOT TEARS IN (B)

Figure 7-31. Hot Tear Discontinuities

(2) Article configuration and metallurgical composition may make demagnetization difficult.

(3) Although magnetic particle testing can detect near surface hot tears, radiography should be used for final analysis.

(4) Foreign material not removed prior to testing will cause an invalid test.

c. Liquid Penetrant Testing Method.

(1) Liquid penetrant testing is recommended for nonferrous cast material.

(2) Method is confined to surface evaluation.

(3) The use of penetrants on castings may act as a contaminant by saturating the porous structure and thereby affect the ability to apply surface finish.

(4) Repeatability of indications may be poor.

d. Ultrasonic Testing Method. Not recommended for detecting hot tears. Discontinuities of this type when associated with cast structure do not lend themselves to ultrasonic testing.

e. Eddy Current Testing Method. Not recommended for detecting hot tears. Metallurgical structure along with the complex configurations do not lend themselves to eddy current testing.

732 INTERGRANULAR CORROSION

1. Category. Service

2. Material. Nonferrous

3. Discontinuity Characteristics

Surface or internal. A series of small micro-openings with no definite pattern. May appear singly or in groups. The insidious nature of intergranular corrosion results from the fact that very little corrosion or corrosion product is visible on the surface. Integranular corrosion may extend in any direction following the grain boundaries of the material. (See Figure 7-32.)

4. Metallurgical Analysis

Two factors that contribute to intergranular corrosion are:

a. Metallurgical structure of the material that is prone to intergranular corrosion such as unstabilized 300 series stainless steel.

b. Improper stress relieving or heat treat may create the susceptibility to intergranular corrosion. Either of these conditions coupled with a corrosive atmosphere will result in intergranular attack.

5. NDT Methods Application and Limitations

a. Liquid Penetrant Testing Method.

(1) Liquid Penetrant testing is the first choice due to the size and location of this type of discontinuity.

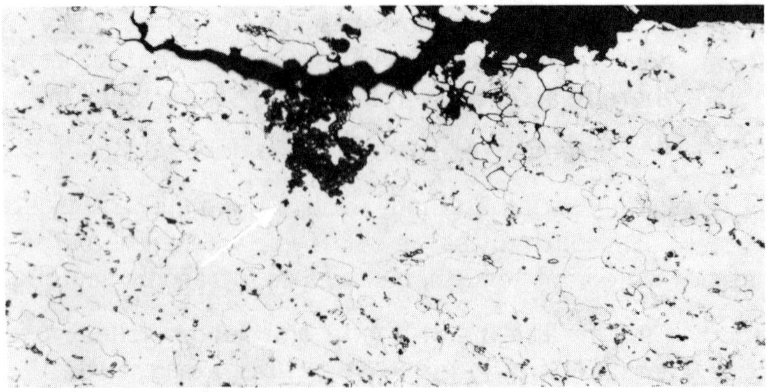

A. MICROGRAPH OF INTERGRANULAR CORROSION SHOWING LIFTING OF SURFACE FROM SUBSURFACE CORROSION

B. MICROGRAPH SHOWING NATURE OF INTERGRANULAR CORROSION. ONLY MINOR EVIDENCE OF CORROSION IS EVIDENT FROM SURFACE

Figure 7-32. Intergranular Corrosion Discontinuity

(2) Chemical cleaning operations immediately before the application of liquid penetrant may contaminate the article and seriously affect test results.

(3) Cleaning with solvents may release chlorine and accelerate intergranular corrosion.

(4) Trapped penetrant solution may present a cleaning or removal problem.

b. Radiographic Testing Method.

(1) Intergranular corrosion in the more advanced stages has been detected with radiography.

(2) Sensitivity levels may prevent the detection of fine intergranular corrosion.

(3) Radiography may not indicate the surface on which the intergranular corrosion occurs.

c. Eddy Current Testing Method.

(1) Eddy current can be used for the screening of intergranular corrosion.

(2) Tube or pipe lend themselves readily to this method of NDT testing.

(3) Metallurgical structure of the material may seriously affect the output indications.

d. Ultrasonic Testing Method. Not normally used although the equipment has the capability to detect intergranular corrosion.

e. Magnetic Particle Testing Method. Not recommended for detecting intergranular corrosion. Type of discontinuity and material restrict the use of magnetic particles.

CHAPTER 8: SAFETY PRECAUTIONS

TABLE OF CONTENTS

CHAPTER 8: SAFETY PRECAUTIONS

800 GENERAL

Liquid penetrant testing uses a variety of materials that have distressing and often hazardous characteristics. Except for water, the liquids used are usually flammable and some, upon contact, can cause skin irritation. The developing powders used are nontoxic but, in confined spaces, can become a health hazard. The black lights used with fluorescent penetrants are in the ultraviolet spectrum of light rays that can cause physiological damage (sunburn, etc.). All of these hazards can be avoided or minimized by observing the precautionary measures mentioned in the following paragraphs.

801 FIRE

Flashpoint is the lowest temperature at which vapors above a volatile combustible substance ignite in air when exposed to flame. Some penetrant materials have very low flashpoints and their use should be avoided. The higher the flashpoint of a material the less fire hazard it presents. Safe practice requires that penetrant materials used in open tanks have a minimum flashpoint of 125°F (52°C). Because most penetrant materials burn readily, smoking is forbidden in or near test areas. Penetrant materials are never stored near heat or open flame, and exhaust fans are used to disperse vapors.

802 SKIN IRRITATION

The oil base of liquid penetrant materials has a drying action on the skin. Because of this, the materials may cause unpleasant, if not dangerous, irritations. To prevent unnecessary contact with penetrant materials care is taken to avoid splashing; protective hand creams are used; aprons, neoprene gloves, and face shields are worn; and soap and water are used to immediately remove any penetrant materials that have come in contact with the skin.

803 AIR POLLUTION

Dust and vapors from materials used in liquid penetrant testing are nontoxic but inhalation of excessive amounts can be a health hazard. To avoid unhealthful concentrations of developer powder in the atmosphere, exhaust fans are installed in confined areas where dry developers are used. Fans are also used to remove the vapors from the test area.

804 BLACK LIGHT

The black light used to cause fluorescence of penetrant materials has a frequency of approximately 3650 angstroms. This frequency is at the low end of the ultraviolet spectrum, the least harmful portion of the spectrum. The higher frequency ultraviolet rays are harmful to many forms of life including humans since they cause sunburn and are injurious to the eyes. The filters used with penetrant test black lights filter out the harmful ultraviolet rays generated by the mercury arc lamp as well as the visible light rays. Missing, cracked, or broken filters are to be replaced before use.

APPENDIX A

GLOSSARY

Angstrom Unit   A unit of length equal to 10^{-8}cm and used to express wavelengths of light; i.e., electromagnetic radiation.

Aqueous Developer   See Wet Developer

Background   The surface upon which an indication is viewed. It may be the natural surface of the test article, or it may be the developer coating on the surface. This background may contain traces of unremoved penetrant (fluorescent or visible), which if present, can interfere with the visibility of indications.

Background Fluorescence   Fluorescent residues observed over the general surface of the test article during fluorescent penetrant inspection.

Bath   Term used colloquially to designate the liquid penetrant inspection materials into which test articles are immersed during inspection process.

Black Light   Light radiation in the near ultraviolet range of wavelengths (3200 to 4000 angstroms), just shorter than visible light.

Black Light Filter   A filter that transmits black light while suppressing visible light and harmful ultraviolet radiation.

Bleedout   The action of the entrapped penetrant in spreading out from surface discontinuities to form an indication.

Blotting   The action of the developer in soaking up penetrant from a surface discontinuity, so as to cause maximum bleedout of penetrant for increased contrast and sensitivity.

Capillary Action   The tendency of liquids to penetrate or migrate into small openings, such as cracks, pits, or fissures.

Carrier Fluid   A fluid in which liquid penetrant inspection materials are dissolved or suspended.

Clean Free of interfering solid or liquid contamination.

Color Contrast Penetrant See Visible Dye Penetrant.

Comparative Test Block An intentionally cracked metal block having two separate but adjacent areas for the application of different penetrants so that a direct comparison of their relative effectiveness can be obtained. Can also be used to evaluate penetrant test techniques and test conditions.

Contact Emulsifier An emulsifier that begins emulsifying penetrant upon simple contact with the penetrant. Usually oil based (lipophilic).

Contrast The difference in visibility (brightness or coloration) between an indication and the surrounding surface.

Dark Adaptation The adjustment of the eyes when one passes from a bright to a darkened area.

Defect A discontinuity that interferes with the usefulness of an article. A fault in any material or part that is detrimental to its serviceability.

Detergent Remover A penetrant remover that is a solution of a detergent in water. Also see Hydrophilic Emulsifier.

Developer A material that is applied to the test article surface after excess penetrant has been removed and that is designed to enhance the penetrant bleedout to form indications. The developer may be a fine powder, a solution that dries to a fine powder, or a suspension (in solvent, water, alcohol, etc.) that dries leaving an absorptive film on the test surface.

Developing Time The time between the application of the developer and the examination of the test article for indications. The elapsed time necessary for the applied developer to bring out indications from penetrant entrapments. Also called Development Time.

Discontinuity An interruption in the normal physical structure or configuration of an article, such as cracks, forging laps, seams, inclusions, porosity, etc. A discontinuity may or may not affect the usefulness of the part.

Dragout The carryout or loss of penetrant materials as a result of their adherence to the articles being processed.

Drain Time That portion of the penetrant inspection process during which the excess penetrant, emulsifier, detergent remover, or developer is allowed to drain from the test article.

Dry Developer A fine dry powder developer that does not employ a carrier fluid.

Drying Oven An oven used for drying rinse water from test articles.

Drying Time The time allotted for a rinsed test article to dry.

Dual Sensitivity Penetrant A penetrant that contains a combination of visible and fluorescent dyes.

Dwell Time The total time that the penetrant or emulsifier is in contact with the test surface, including the time required for application and the drain time. Also see Emulsification Time.

Electrostatic Spraying A technique of spraying wherein the material being sprayed is given a high electrical charge, while the test article is grounded.

Emulsification Time The period of time that an emulsifier is permitted to combine with the penetrant prior to removal. Also called Emulsifier Dwell Time.

Emulsifier A liquid that combines with an oily penetrant to make the penetrant water washable. Also see Hydrophilic-Emulsifier and Lipophilic Emulsifier.

Evaluation The process of determining the severity of the condition after the indication has been interpreted. Evaluation leads to determining whether the article is acceptable, salvageable, or rejectable.

False Indication An indication due to improper processing. Distinct from nonrelevant indication.

Family The complete series of materials necessary to perform a specific process of liquid penetrant inspection.

Flash Point The lowest temperature at which a volatile flammable liquid will give off enough vapor to make a combustible explosive mixture in the air space surrounding the liquid surface.

Flaw See Defect.

Fluorescence The emission of visible radiation by a substance as the result of, and only during, the absorption of black light radiation.

Fluorescent Penetrant An inspection penetrant that is characterized by its ability to fluoresce when excited by black light.

Hydrophilic Emulsifier A water-based agent that, when applied to an oily penetrant, renders the penetrant water washable. Can be used as a Contact Emulsifier, but more often the emulsifier is added to the water rinse and accompanied by some form of mechanical agitation or scrubbing to remove excess penetrant. Sometimes called a Hydrophilic Scrubber.

Indication That which marks the presence of a discontinuity, as the result of detectable bleedout of penetrant from the discontinuity.

Inspection The visual examination of a test article after completion of the penetrant processing steps.

Interpretation The determination of the significance of indications from the standpoint of whether they are relevant or nonrelevant.

Leak Testing A technique of liquid penetrant testing in which the penetrant is applied to one side of the surface, while the other side is inspected for indications that would indicate a through-leak or void.

Lipophilic Emulsifier An oil-based agent that, when applied to an oily penetrant, renders the penetrant water washable. Usually applied as a Contact Emulsifier.

Nonaqueous Wet Developer A developer in which the developing powder is applied as a suspension in a quick-drying solvent. Also called Solvent Developer.

Nonfluorescent Penetrant See Visible Dye Penetrant.

Nonrelevant Indication An indication that is not or cannot be associated with a discontinuity.

Penetrability The property of a penetrant that causes it to find its way into very fine openings, such as cracks.

Penetrant A liquid (sometimes gas) capable of entering discontinuities open to the surface, and which is adapted to the inspection process by being made highly visible in small traces. Fluorescent penetrants fluoresce brightly under black light while the visible penetrants are intensely colored to be noticeable under visible light.

Penetration Time See Dwell Time.

Post-Emulsification A penetrant removal technique employing a separate emulsifier.

Post-Emulsification Penetrant A penetrant that requires the application of a separate emulsifier to render the surface penetrant water washable. Also can be removed by applying a solvent remover.

Precleaning The removal of surface contaminants or smeared metal from the test article so that they cannot interfere with the penetrant inspection process.

Quenching of Fluorescence The extinction of fluorescence by causes other than removal of black light (the exciting radiation).

Rinse The process of removing liquid penetrant inspection materials from the surface of an article by washing or flooding with another liquid — usually water. Also called Wash.

Seeability The characteristic of an indication that enables the observer to see it against the conditions of background, outside light, etc.

Self-Emulsifiable See Water Washable.

Sensitivity The ability of the penetrant process to detect surface discontinuities.

Solvent Developer See Nonaqueous Wet Developer.

Solvent Removed A penetrant removal technique wherein the excess penetrant is washed or wiped from the test surface with a solvent remover.

Solvent Remover A volatile liquid used to remove excess surface penetrant from the test article. Sometimes called Penetrant Remover.

Surface Tension That property of liquids which, due to molecular forces, tends to bring the contained volume into a form having the least superficial area.

Viscosity The state or degree of being viscous. The resistance of a fluid to the motion of its particles.

Visible Dye Penetrant An inspection penetrant that is characterized by its intense visible color - usually red. Also called Color Contrast or Nonfluorescent Penetrant.

Wash See Rinse.

Water-Soluble Developer A developer in which the developer powder is dissolved in a water carrier to form a solution. Not a suspension.

Water-Suspended Particle Developer A developer in which the developer particles are mixed with water to form a suspension.

Water-Wash A penetrant removal technique wherein excess penetrant is washed or flushed from the test surface with water.

Water-Washable Penetrant A type of penetrant that contains its own emulsifier, making it water washable.

Water Tolerance The amount of water that a penetrant, emulsifier, or wet developer can absorb before its effectiveness is impaired.

Wet Developer A developer in which the developer powder is applied as a suspension or solution in a liquid - usually water or alcohol.

Wetting Ability The ability of a liquid to spread out spontaneously and adhere to solid surfaces.

APPENDIX B

MEASUREMENT CONVERSION CHARTS

U.S. CUSTOMARY TO INTERNATIONAL SYSTEM (SI) UNITS

Inch to Millimeter Conversions							
Inch	Millimeter	Inch	Millimeter	Inch	Millimeter	Inch	Millimeter
1	25.4	26	660.4	51	1295.4	76	1930.4
2	50.8	27	685.8	52	1320.8	77	1955.8
3	76.2	28	711.2	53	1346.2	78	1981.2
4	101.6	29	736.6	54	1371.6	79	2006.6
5	127.0	30	762.0	55	1397.0	80	2032.0
6	152.4	31	787.4	56	1422.4	81	2057.4
7	177.8	32	812.8	57	1447.8	82	2082.8
8	203.2	33	838.2	58	1473.2	83	2108.2
9	228.6	34	863.6	59	1498.6	84	2133.6
10	254.0	35	889.0	60	1524.0	85	2159.0
11	279.4	36	914.4	61	1549.4	86	2184.4
12	304.8	37	939.8	62	1574.8	87	2209.8
13	330.2	38	965.2	63	1600.2	88	2235.2
14	355.6	39	990.6	64	1625.6	89	2260.6
15	381.0	40	1016.0	65	1651.0	90	2286.0
16	406.4	41	1041.4	66	1676.4	91	2311.4
17	431.8	42	1066.8	67	1701.8	92	2336.8
18	457.2	43	1092.2	68	1727.2	93	2362.2
19	482.6	44	1117.6	69	1752.6	94	2387.6
20	508.0	45	1143.0	70	1778.0	95	2413.0
21	533.4	46	1168.4	71	1803.4	96	2438.4
22	558.8	47	1193.8	72	1828.8	97	2463.8
23	584.2	48	1219.2	73	1854.2	98	2489.2
24	609.6	49	1244.6	74	1879.6	99	2514.6
25	635.0	50	1270.0	75	1905.0	100	2540.0

To convert millimeters to contimeters, move decimal point one place to the left (e.g., 635 mm = 63.5 cm)

To convert millimeters to meters, move decimal point three (3) places to the left (e.g., 635 mm = 0.635m)

Common Fractions to Millimeter Conversions	
Inch	Millimeters
1/64	0.397
1/32	0.794
1/16	1.588
1/8	3.175
1/4	6.350
1/2	12.700
1	25.400

NOTES

NOTES

NOTES

NOTES

NOTES

NOTES